NO GRID SURVIVAL PROJECTS BIBLE

Build Your Self-Sustainable Oasis with Recession-Proof DIY Projects and Prepper's Alpha Techniques. House Protection, Endless Food, Water, Energy Supply & Beyond

CARTER FIELDSTONE

TABLE OF CONTENTS

INTRODUCTION

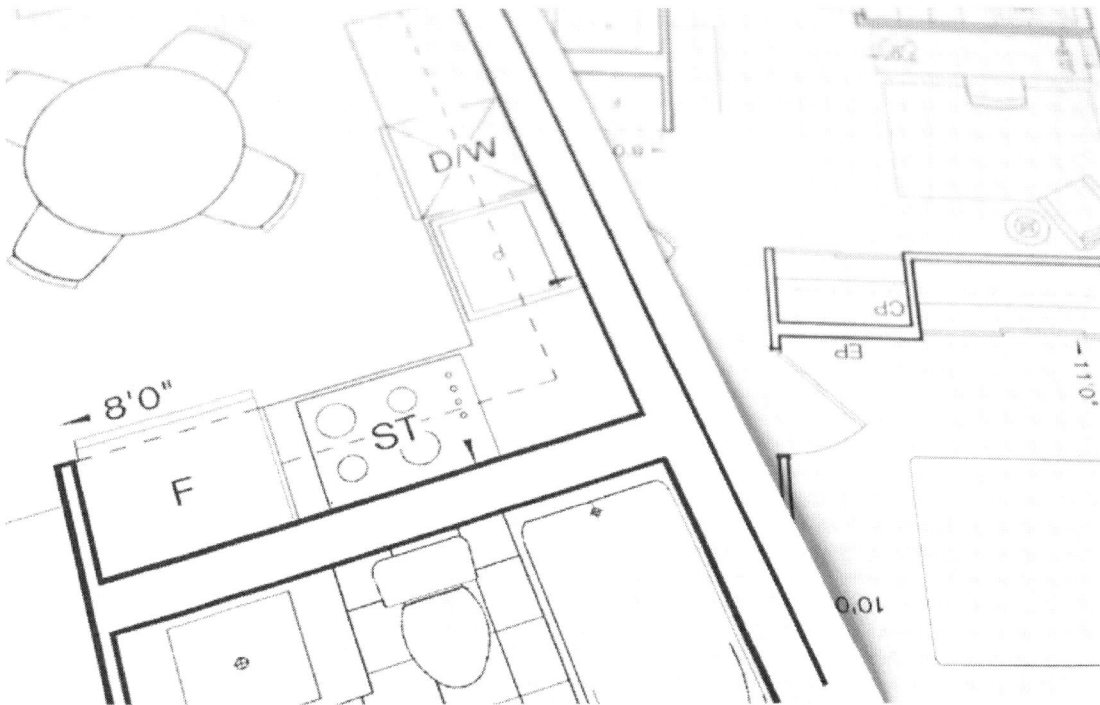

Just take a minute to imagine a life free from being tied down to the grid. A life where you are in complete control and can define what self-sufficiency and resilience entail. Welcome to the no-grid community, an idea that has remained under the radar but is slowly regaining traction across the United States.

Let me stop you right here before you conjure up images of isolated cabins in the woods with a pet grizzly. No-grid living isn't a camouflage for going off the deep end and escaping society; it's a new definition of the resources we depend on as well as the communities we live in. Americans have thrived on no-grid living (even if it wasn't called this) for ages, not only by demonstrating a certain sense of independence and an "on-the-go attitude" but also a deep respect for the land. From the early pioneers who carved out a life on the frontier to the groovy back-to-the-land movement of the 1960s and 70s who wanted to stick it to the man, the desire to live life on our own terms has always been a part of

American culture and spirit. But here's the current state of affairs: no-grid or off-the-grid living isn't just for the core doomsday homesteads or gritty survivors anymore. In recent years, it's become a mainstream consideration and has attracted not only people who have chosen to live a green way of life but also individuals from all walks of life.

On the other hand, there is the question of the philosophy of the no-grid way of living. It's about realizing that the present grid-based systems are fragile and unsustainable, so we should take on our own ability to be prepared for any situation. In a nutshell, it means living a life where you have a mindset that fosters self-reliance, being environmentally friendly, and staying actively prepared. You don't only save yourself, but you also contribute towards a more sustainable environment when you choose the no-grid lifestyle.

However, no-grid existence is not just a matter of being a solo wolf. It's about creating a supportive community and working together. It is about organizing mobilization towards the creation of friendship circles that educate and uplift one another. It is a matter of spreading education and resources to create a world where everyone is linked and stable. This book right here will tell you everything about this very precise situation. What's coming up next is exploring how you can transition to a gridless life with a wealth of resources and practical ideas to help you get started. We'll be going deeper into the reason why people aim for self-sufficiency, unpacking the ethics and practicalities behind disconnecting from grid-based lifestyles. We will also discuss off-grid variants in further detail, addressing the possible problems you can face during the transition, and providing you with the necessary roadmap to navigate this new phase of your life successfully.

We must not limit ourselves there. This book is a complete self-sufficiency manual enriched with step-by-step guides and basic skills to thrive in the no-grid world. We'll be talking about how to collect and clean water as well as about growing food by gardening and raising animals. We will additionally be covering renewable energy options for your home which will enable you to live your daily life using clean green energy for your electricity.

The obvious truth is you cannot just live off the grid in terms of pragmatism. It also focuses on strengthening your home while also facilitating the safety and security of your family. We will walk through the most important security and shelter measures, as well as the emergency preparation skills and kits you should have on hand to make sure you are well prepared for anything.

And tending to things could also be fun. This book is the ultimate DIY off-grid project source, which will not only teach you valuable skills but also give you fun and satisfaction. From creating your own solar oven to making a warm and comfortable one without electricity, there are things for everyone. Considering both the seasonal and beginner no-gridders, that is your most decisive hand guide. Prepare yourself for a story of overcoming adversity, conservation, and self-reliance.

CHAPTER 1
WHY SELF-SUFFICIENCY

So the unfamiliar greenhorns amongst you hear "no-grid living" and immediately envision an encampment of wild-eyed doomsayers stockpiling canned beans and sharpening hatchets, right? You might want to forget those dried-up misconceptions from the musty cupboards of your mind.

At its core, the heartbeat of this movement stems from a sincere desire to rekindle the flames of self-reliance, resilience, and a solid dose of good old-fashioned adaptability. For too long, we've been conditioned to blindly entrust our sustenance and security to increasingly fragile systems that leave us at the mercy of powers beyond our control.

Every power outage, supply chain disruption, and injustice of centralized authority is a blaring wake-up call to take back control of our own lives.

What is Self-Sufficiency?

As we stand on the precipice of your no-grid life, you're going to need to know a little more about the fundamental philosophy that will guide you toward a life of self-sufficiency.

Self-sufficiency is the instinctive human desire for independence, self-reliance, and personal control over your daily needs. To be truly self-sufficient means having the capabilities and resources to take care of your basic needs without relying on outside authorities, public services, or fragile systems. Now make no mistake, this seemingly rugged ideal doesn't necessarily relegate you to a primitive existence where you stroll off foraging for grub and skinning game in some dank cave crevice. Even in this modern world of ours, you can still embody that pioneering spirit in your own unique way.

For some off-grid pioneers, self-sufficiency may mean cutting the cord entirely from public utilities like power grids and municipal water sources. Imagine your very own off-grid oasis, where you're generating your own clean, green energy, capturing pure rainwater, cultivating bountiful hormone-free garden harvests, and remaining independent from external authorities in ivory towers.

Maybe you're not quite ready to go all-in on the homesteading life. That's perfectly okay. You can still embrace the "no-grid" existence while living in the heart of the city. Stay hooked up to the grid when it's working, but have a solid plan B in your back pocket for when those systems hiccups or crises inevitably strike. Just look at the chaos of the past few years; empty shelves, rolling blackouts, you name it. Having a self-sufficient mindset is just plain smart living.

Whether your personal vision leans towards an off-grid paradise or a more no-grid integration, the core principles remain the same. Self-sufficiency boils down to an unbreakable, can-do mentality of industrious problem-solving, versatility, and resilience in the face of any challenge. It's about nurturing capabilities to reduce dependence on external sources by producing more for yourself through:

- Cultivating your own vitamin-rich fruits, vegetables, and livestock.

- Harnessing clean renewable power sources like solar, wind, etc.

- Collecting, purifying, and managing your own water supply.

- Learning vital hands-on skills to repair, build, and craft necessities.

- Adopting eco-conscious practices to reduce waste and live lightly.

By developing these empowering competencies, you build strong personal defenses against external vulnerabilities, weaknesses, and unpredictability. In a self-sufficient home, you're less affected by resource shortages, power outages, and economic instability that would otherwise disrupt daily life. But understand this truth; self-sufficiency isn't just about "surviving" some apocalyptic doomsday event. It is about living in harmony with Mother Nature, respecting the land that sustains us, and breaking free from the cycle of mindless consumption that's been drilled into our heads. When you embrace self-sufficiency, you're not just claiming your independence; you're reconnecting with the earth in a profound and meaningful way.

So call it off-grid autonomy, no-grid sensibility, or self-reliance, the name itself carries little importance. What matters most is the mindset. It rejects the helplessness and dependency that society tries to sell us and instead, embraces the empowering self-actualization of personal resilience. It combines sturdy sustainability with environmental harmony.

Analyzing Current Environmental Challenges

Self-sufficiency means creating personal independence and resilience through developing crucial capabilities: skills that empower you to meet your core needs without being tied down to increasingly fragile, centralized systems.

But the mission goes way beyond just being self-reliant. The very justification for embracing the no-grid revolution is rooted in a calling so noble, it's practically divine. One that's rooted in our obligation as caretakers and defenders of this planet we call home.

You just need to take a look at the environmental catastrophes unfolding across our planet. The climate crisis is accelerating with every reckless emission of greenhouse gasses trapping heat. Ice caps are thawing away, sea levels are consuming coastlines, and weather patterns are morphing into Biblical-scale extremes. Heatwaves, megadroughts, and supercharged storms are all evidence of Earth lashing back.

Even those who firmly deny it cannot overlook the rapidly decreasing supply of limited resources like fossil fuels, clean water, fertile soil, and essential minerals that our industrial societies have mercilessly plundered for their pockets. We're gazing into an abyss of scarcity if we maintain these outrageously, unsustainable practices.

What about the plague of pollution, the toxic byproduct of humanity's wasteful arrogance? Colossal islands of garbage polluting our oceans. Air is choked with smog and carcinogens. Chemicals and heavy metals seep into the soils and waterways that nourish all life on this planet.

You simply cannot turn a blind eye or shrug your shoulders in ignorance of these existential emergencies. The alarm sounds very clear through every scientific update, every noticeable shift in the climate, and every new water or food problem. Though this bleak overview may seem like the beginning of civilization's decline, it also exposes the significant purpose behind the off-grid movement that shines brighter than a solar panel at high noon.

Embracing self-sufficiency and the no-grid lifestyle initiates an unavoidable reconnection to the natural world, pioneering a balanced approach to taking care of our environment. When you grow your own organic crops and raise livestock through permaculture and regenerative practices, you nurture the soil's health instead of depleting it. You eliminate the pollution caused by emissions from transporting store-bought items over long distances. When you harvest rainwater and implement greywater recycling systems, you reduce strain on rapidly dwindling freshwater reserves and minimize toxic runoff. You trim reliance on fossil-fueled utilities by transitioning to renewable off-grid power like solar, wind, geothermal, or micro-hydro, combating greenhouse gas emissions.

At every turn, the no-grid ethos requires adopting eco-conscious habits like composting, natural building with sustainable materials, and shrinking your consumption footprint. You'll gain insights into the interconnected natural cycles sustaining all life, fostering a deep sense of gratitude for life itself.

You will become more than just a bunch of self-reliant homesteaders living off the fat of the land. You become part of an influential movement nudging society towards that crucial philosophical shift; away from egocentric domination over nature towards a more harmonious way of living. Instead of trying to control everything, you will be focusing on working with nature, sharing resources fairly, and making sure everyone has what they need to thrive.

By restoring your personal independence and resilient self-sufficiency, you're also working wonders to heal and safeguard this ailing planet that nurtures us all.

The Benefits of No-Grid Living

We've shone a spotlight on how wasteful and selfish human behavior is wreaking havoc on the planet. At the same time, we've also highlighted how reconnecting with sustainable living can bring about positive change.

Now, I'm not going to deny that no-grid living is completely altruistic. Even if you're motivated by spiritual fulfillment you're going to find yourself craving tangible, real-world rewards for all your blood, sweat, and tears from living off the grid. It's natural to want more than philosophical satisfaction. Well, steady your roots, because the benefits of no-grid living go beyond just ecological redemption and taking care of nature.

- Enough is enough when it comes to pouring your hard-earned cash straight into those corporate coffers while keeping you hopelessly chained to their grids. It's high time you liberated yourself and tasted the sustaining nectar of true self-reliance. That's why pursuing an off-grid, no-grid lifestyle isn't just some tree-hugging pipe dream. It's a legitimate path to economic emancipation and personal independence over your resources. A bold "stick it to the man" move that will pay off for decades. Here's how it works in short:

> » Put in the upfront investments to construct your own self-sustaining homestead haven. Install solar panels on your roof to bask in renewable power and independence.

> » Drill a well or set up atmospheric water capture to create self-replenishing freshwater reserves.

> » Implement greywater recycling and waterless composting to intelligently close the waste loop while nurturing your fertile land.

Yes, it requires some initial financial willpower. However, these investments lead to exceptional personal freedom and save money in the long run. It's a solid investment. No more endless utility charges lining the pockets of utility gougers.

- The self-sufficient path doesn't stop at just cutting the municipal utility cord. For the truly devoted homesteaders, it means fully immersing into a life of permaculture food forests and nurturing livestock. Can you imagine the reward of rediscovering unmatched flavors, nutrition, and soul-nourishing pride in cultivating your own raw provisions?

  > » No more mindless mainstream meals getting you nickeled-and-dimed for watery, pathetic grocery commodities shipped from who-knows-where.

  > » You'll be savoring vitamin-dense harvests. Pure, wholesome, untainted nourishment from your own labor and harmony with the land.

- Your entire spirit gets replenished when you see the tangible results of self-empowered self-reliance. With each step deeper into this lifestyle, confidence and problem-solving grit are fostered at impeccable levels.

  > » Gone are the days of helpless dependence on faceless corporations and bureaucrats who could care less about your well-being.

» You're more than a number, you're a self-sufficient person.

» In turn, this self-sufficiency brings the comforting confidence that your abilities and strength can endure any crisis or disruption.

It's not all a downhill freeroll into Easy Street. Mastering homesteading arts like livestock farming, seed saving, food preservation, and natural building will stretch your mind, imagination, and capabilities at first. Learning how to manage water sustainably, craft tools by hand, and adapt to your local climate are all practical and hands-on skills. But isn't that what real personal growth is all about? Each new challenge is just an opportunity to expand your skills and resistance. And for all the doubters brooding "What about my precious conveniences and modern creature comforts?" Many of today's off-grid innovations integrate self-sufficiency with desired comforts when used together effectively.

From cost-efficient solar appliances and robust home battery storage to global satellite internet like Starlink, you can craft your ideal "off-grid luxury" experience. Don't buy into those limiting myths that living off-grid means living in hardship. Will it be a cakewalk every single day?

No. Overhauling your entire lifestyle into self-sufficient sustainability demands stamina. You'll be living a self-reliant life by carefully maintaining systems, fixing parts, and keeping an eye on resources like water reserves. But stare those roadblocks straight in the eye and recognize them for what they are: Mindset-strengthening beginnings.

CHAPTER 2
THE TRANSITION TO OFF-GRID LIVING

So, you've awoken to the galvanizing potential of the no-grid life, have you? Are you feeling that itch to break free from the shackles of unsustainable consumption and embrace resilient self-sufficiency?

Modern society has cocooned us in what feels like an inescapable web of complete dependence on unsustainable grid systems. A sluggish existence where we're bound to utility overlords for every single basic necessity like powering our homes, hydrating our bodies, and even putting food on the table. Moving away from restrictive ties and shifting to an independent off-grid lifestyle won't magically happen through simply philosophizing and daydreaming about it. Achieving self-reliance requires a comprehensive overhaul of how you provision every single resource required to not just survive, but thrive; energy, water, food, and even the roof over your head.

Before you start sweating buckets over the magnitude of this undertaking, take a few deep breaths. This eye-opening chapter is going to walk you through the process of building your self-sustaining domain, step by step. We'll help you identify all the key areas that require transformation and give you a foolproof roadmap for adopting integrated no-grid systems.

Look, I'm not going to sugarcoat it; building a life free from grid dependency is going to take a whole lot of patience, determination, and an insatiably curious mind. There will be inevitable setbacks and challenges testing every last fiber of your newfound self-reliant grit. But keep your eye on the prize, because the fruits of off-grid existence are absolutely mind-blowing.

Factors to Consider Before Transitioning

Before you go charging off into the wild, you need a grounding pause for some contemplation. As tantalizing as this transformative process seems, with its promises of spiritual growth and eco-friendly living, the unfiltered truth is that it's a bold move that comes with its fair share of practical challenges.

Not everyone's unique living situation is going to be a perfect fit for diving headfirst into the no-grid lifestyle right off the bat. There's an entire checklist of practical prerequisites and logistical considerations that must be carefully weighed before you can responsibly start untangling yourself from the grid.

A little bit of foresight and self-awareness now will save you a world of headaches and wasted resources down the line. So, let's put those starry-eyed visions of off-grid paradise on hold for just a minute. It's time to take an honest, no-nonsense look at these essential factors.

- Location, Location, Location

The very foundations of your dream off-grid life hinge on finding the perfect spot for your self-sufficient home base. There's a lot more to it than just hunting down a pretty place! You have to take a good, hard look at aspects like:

- Regional climate patterns

  » Availability of renewable energy sources

  » Soil fertility for high-yield gardening ops.

  » Freshwater access

  » Relative remoteness from commercial services and healthcare infrastructure

For some, slowly integrating off-grid tech and practices while still being close to the city may be optimal. But, for the big dreamers wanting a full-fledged rural oasis, some serious homework needs to be done to pinpoint areas that can truly maximize off-grid self-sufficiency potential.

- Home Construction Realities

Once you've zeroed in on a promising location, the next consideration is whether your existing residential situation can realistically support all the retrofitting and system upgrades you'll need to make. Or, is it time to start from scratch and build the resilient homestead of your dreams?

If you're starting from ground zero with new construction, get ready to dive into the world of alternative building designs. This will involve factors like:

- Blending modern sustainable innovations with traditional ancestral techniques

  » Passive solar design for natural climate control

  » Use of eco-friendly construction materials

  » Accommodating renewable energy installments

Every choice you make will either boost or bust your self-sufficiency goals, so choose wisely!

- Getting Self-Sufficient at Climate Control

Your shelter's thermal envelope and temperature regulation capabilities simply have to emerge as absolutely mission-critical priorities.

Any slack on things like proper passive solar design, insulation, or strategic landscaping and exterior shading will undermine your heating/cooling self-sufficiency in hugely compromising ways.

Essentially, you'll be draining way too much precious energy from your solar panels or generator banks just to keep yourself comfortable. Conducting exhaustive property and site assessments to figure out the best way to integrate passive conditioning is non-negotiable.

- Power to the Self-Reliant People

The next step involves intelligently designing your homestead's off-grid electrical integration. But before you start throwing solar panels on your roof all willy-nilly, you need a solid handle on just how much energy your household is going to need.

Every single calculation you make—from figuring out how many solar panels you need, whether you want to add in some wind or micro-hydro generators, to how much battery backup you should have—depends first on gaining insight into your electrical load equation.

What energy-intensive appliances and equipment like well pumps, medical devices, or power tools need accommodating? How smoothly can you transition from fossil-fueled generators to renewable storage solutions? Your electrical self-sufficiency blueprint is not the time to wing it!

- Water Autonomy or Bust

Access to reliable freshwater supplies arguably stands as the most important resource consideration when you're going off-grid.

You'll need to do some serious homework for optimal atmospheric humidity harvesting on:

- Groundwater surveys

  » Surface runoff seasonality patterns

  » Hydrology assessments

» Rain catchment surface calculations

» Storage tank capacities

Mapping out your water distribution network and mastering your hydrological systems is key to achieving true water independence. Ignoring any aspects of this at your potential homestead is a one-way ticket to Struggle City.

- Solving the Sanitation Equation

Grid-free households can no longer rely upon those centralized municipal sewage conveyance systems once ties are cut from surrounding infrastructure.

Strategies for sustainably recycling greywater streams and establishing self-contained blackwater treatment solutions require planning and implementation. Options worth exploring include:

- Permaculture-inspired constructed wetland bio-filters

  » Living machine detoxifiers

  » Waterless composting toilet setups designed to recycle nutrient-rich fertilizer

Pick whichever regenerative approach best aligns with your long-term self-sufficiency goals.

- Cultivating Total Food Autonomy

No off-grid existence can ever hope to thrive if you're still relying on industrial supply chains for your food.

To generate abundant dietary staples for your household, you have to start implementing intelligent, high-yield permaculture practices. You'll need to analyze what works best for your local microclimate patterns, solar profile, and cultivable acreage. Learn and get comfortable with:

- Planning diverse crops throughout the seasons

  » Composting and intensive gardening methods

  » Taking care of the soil in a regenerative way

» Including resilient livestock and beekeeping in your food production system

It might seem daunting at first, but there's nothing more empowering than being able to feed yourself and your loved ones with your own two hands.

- Tips

The truth here? Going 100% off-grid and breaking free from the shackles of consumer culture is a complete life overhaul. This is not some hobbyist-level project.

For so many off-grid truth-seekers just starting to wake up to the call of self-reliance, the smartest way forward is often a gradual, step-by-step approach over the course of a few years rather than weeks or months.

- Begin by methodically integrating off-grid technologies, practices, and permaculture principles in conjunction with one foot in the world of municipal utilities and services.

  » Get your feet wet first with some urban homesteading, installing renewable power, and water catchment. Run some self-sufficiency "dry run" tests while still having a safety net.

  » Only when you've got your core systems running like a well-oiled machine and you've proven you can handle intermediate self-reliance should you even think about cutting the cord completely.

This gradualist approach trades the fantasy of overnight radical autonomy for the reality of gradually building skills and minimizing risks. It's all about being practical and patient, not reckless and delusional. But for the uncompromising few, then by all means, charge forward with laser focus and unshakable determination. Just make sure you've got the physical, mental, and adaptability chops to handle the massive upheaval headed your way.

Economic Considerations

Even with all the incredible empowerment and soul-enriching benefits of being truly self-sufficient, one issue often holds people back—"How can I possibly finance such a comprehensive overhaul away from public utility dependence? Isn't constructing a fully self-reliant off-grid homestead prohibitively expensive for most?"

Well, calm your economic anxieties. The reality remains that cutting ties with the industrial world means making some strategic investments in the right tech and infrastructure. But, if you're smart with your cash, there are tons of ways to keep those surprise expenses in check. Let's start by looking at the two primary expenditure fronts you'll face:

- Upfront transition costs: This entails construction materials for resilient shelters, renewable hardware like solar arrays and wind turbines, water autonomy solutions, and high-yield foodscaping overhauls to name but a few immediate requirements. It's a hefty chunk of change, but it's a one-time deal. And here's the beautiful thing about a lot of off-grid tech: once you've got it, you're not constantly hemorrhaging money on bills. Imagine the ballooning utility bills and greedy rent-seeking you'll be sidestepping once your solar-powered energy independence is achieved! No more cash bleeding for water, electricity, or heating from monopolists. Just your very own self-sustaining kingdom.

- Operating costs: The second economic consideration is your ongoing costs for consumables, maintenance, and keeping an eye on your integrated systems. Propane for generators and appliances, parts replacements, soil amendments, and supplemental provisions are all examples of extra provisions you might need. With some practical planning, these expenses become a tiny fraction of what you used to pay for grid-based living.

How does the savvy self-sufficiency pioneer tackle initial setup challenges while steadily bringing their off-grid vision to life? The smartest approach is usually a combo of two strategies:

Transition Incrementally

For a lot of folks, gradually rolling things out over a few years is the way to go. It minimizes capital shocks while still making progress towards your ultimate goal of total autonomy.

Urban or suburban "grid-zens" should start by establishing some foundational permaculture food gardens, rainwater catchment systems, composting setups, and a modest solar array to offset your utility bills.

Rome wasn't built in a day! Each step of the process is a chance to learn, refine your resource management skills, and become more self-reliant. Slow and steady wins the race by avoiding reckless overcommitments while building confidence and familiarity. Only when you've achieved benchmarks for renewable power, water self-sufficiency, and lifestyle adaption, should you even think about committing further to grid-free living

Leverage Project Economics

Even those starting small with rural living can be smart about staging off-grid projects by crunching numbers and being realistic about costs.

Doing a thorough cost-benefit analysis over the entire lifespan of your off-grid setup will help you focus on the most efficient approaches that give you quick returns and keep spending in check. For example:

- Investing in solar panels upfront for energy independence.

- Decentralizing household water supply through installing a well or atmospheric harvesting setup.

- Investing early in establishing high-yield permaculture food forests and edible guilds.

At the end of the day, it's all about prioritizing and phasing in those big-ticket, off-grid systems steadily over time using strategic financing strategies. Maybe you kick it off by tapping governmental clean energy rebates or joining a peer-to-peer microlending cooperative within your local community to cover material costs.

As those first renewable utilities, like solar arrays and water catchments, start bankrolling savings from offsetting utility charges, you can then funnel those savings into funding the next self-reliant system upgrade. In no time a beautiful positive feedback loop manifests where each sustainable investment helps underwrite the next.

Don't sleep on getting scrappy with reclaiming and reusing quality equipment either. If you've got the skills to negotiate good deals on batteries, hardware, construction materials, and more through regional trade networks, even the most lavish self-reliant sanctuaries can be surprisingly affordable right from the start.

And for all my urbanite friends who can't completely cut ties with municipal services, you can still start deploying hybrid off-grid solutions around your home, like solar water heaters, greywater recycling for your gardens, and thermal batteries for power outages. Every step taken reduces dependence on the grid while buying you peace of mind and shrinking your carbon footprint.

Beyond just crunching numbers on infrastructural investments, there are those niggling worries about the daily lifestyle economics too. Like this anxious voice whispering that self-reliance must mean living some kind of bare-bones, monk-like existence of subsistence farming or whatever. Those are outdated misconceptions. You can blend comprehensive off-grid self-sufficiency with modern comforts these days

- Start by establishing lush permaculture food guilds integrating compact livestock management like chickens where possible.

- Use food preservation methods to bulk-process your harvests for year-round stockpiles. The only thing you'll be saying goodbye to is exorbitant grocery bills, not your modern palate.

- Feeling inspired? Sustainably clothe and accessorize yourselves through reviving heritage textile arts like weaving, natural dyeing, felting, and crafting with leather.

- Heat your shelter efficiently through optimized passive solar architecture combined with integrated thermal batteries and smart climate controls.

There's an economically empowered answer to meeting every modern need through repurposing, reusing, or locally producing it yourself with a bit of applied creativity and artisan spirit. Solutions like reusing furniture creatively and organizing bulk purchases cooperatively have already taken off among intentional communities.

Speaking of community. Tap into regenerative neighborhood-sharing economies and ethical barter networks as well. These make it easy to combine surplus yields from households or specialized goods from local producers. Models like these make prosperity feel natural again, rather than forcing you into some kind of needless austerity.

Ultimately, transitioning to a no-grid autonomy mindset does demand carefully reassessing which so-called "essentials" you've been conditioned to believe truly enrich your life versus consumerist addictions robbing you of purpose. With some strategic planning and a willingness to get creatively scrappy, the long-term economic freedom is within reach.

The watchwords? Adaptability, patience, creativity, and heeding the liberating call of true economic self-determination!

The Mental Transition Shift

You can have all the nuts, bolts, and positive cash flow in the world, but there's one crucial piece of your self-reliant journey that you can't just calculate or buy at the store. An intangible X-factor that's more elusive yet more empowering than any eco-village blueprint or equipment stockpile: The powerful inner transformation that drives your shift to off-grid living.

It's a profound change in mindset and spirit that can't be confined by physical structures. It's about cultivating an entirely new self that embraces growth, determination, and passion.

The transition from being a helpless, grid-dependent, industrial servant to living a thriving life of holistic autonomy means rejecting the illusion of constant comfort and convenience that society often tries to sell us. The no-grid movement demands a steadfast willingness to reprogram conditioned expectations and emotional responses from the ground up. Being uncompromisingly open to constant adaptation and facing the unknown head-on becomes part of your very DNA.

As much as this undertaking is about the logistical dominos of setting up systems, at its core, it's a complete rejection of the consumption-based conditioning that shackles modern minds. Generations have been tricked into complacency and disconnected helplessness by paradigms that promote passivity. But not those who dare to expand their awareness beyond mindless entertainment and excessive overthinking. To empower self-sufficiency, you must focus on developing a strong sense of inner control. It begins with nurturing a growth mentality 24/7. It's the unshakable belief that you can steadily master any new skill, no matter how daunting, through dedication and resourcefulness. Roadblock after obstacle should be creatively dismantled and reframed as a chance to upskill.

Any essential craft from sustainable forage gardening and mushroom cultivating to carpentry, blacksmithing, and wildcrafting herbal medicines will stretch your capabilities. Embrace that prospect! Welcome every obstacle that demands greater adaptability, problem-solving, and personal development. And don't doubt for a second that deep-rooted mental patterns, years in the making, will put up a fight against this evolution.

Fears and cyclical negativities about discomfort, sacrifice, and hazard will attempt to sabotage your momentum. Dissolve those illusions by keeping your eyes on the bigger picture. Because the truth lies in recognizing that every moment of growth compounds exponentially into self-empowerment.

What initially manifests as failing, getting filthy, or tweaking a wonky installation inevitably transforms into immense pride and self-efficacy. Remember, mastery is the fruit of stumbling just a bit less each day. From harvesting and naturally preserving nutrient-dense provisions to weaving your own garments from homespun textiles, true independence will become as intuitive as breathing itself!

So embrace the flows and reactivate the dormant circuitry of your natural learner instincts to undergo the profound internal shift that will enable you to thrive in any environment.

- Transitioning Checklist

For convenient reference, this checklist covers essential considerations before transitioning to off-grid living.

Factors to Consider Before Transitioning	Examples	Comments/ Additional Notes
Location	Regional climate patterns like hot and arid, cold and snowy, or mild and temperate climates	
	Availability of renewable energy sources such as solar, wind, or hydroelectric power	
	Soil fertility for high-yield gardening ops like testing soil pH, nutrient levels, and organic matter content	

Factors to Consider Before Transitioning	Examples	Comments/ Additional Notes
Location	Freshwater access including wells, springs, or rainwater catchment systems	
	Relative remoteness from commercial services and healthcare infrastructure	
Home Construction	Blending modern sustainable innovations with traditional ancestral techniques like cob or straw bale homes	
	Passive solar design for natural climate control like south-facing windows, thermal mass, and overhangs	
	Use of eco-friendly construction materials such as recycled wood, bamboo, or adobe bricks	
	Accommodating renewable energy installments like solar panels, wind turbines, or micro-hydro generators	

Factors to Consider Before Transitioning	Examples	Comments/ Additional Notes
Climate Control	Proper passive solar design, insulation, and strategic landscaping for heat retention and cooling	
Self-Reliance	Understanding your electrical load equation by listing all appliances and their energy consumption	
	Intelligently designing your homestead's off-grid electrical integration for efficient energy usage	
Water Autonomy	Atmospheric humidity harvesting through dehumidifiers or dew collectors	
	Groundwater surveys for well placement and depth	
Water Autonomy	Hydrology assessments to understand surface water flow and potential catchment areas	
	Rain catchment surface calculations based on roof area and rainfall patterns	
	Storage tank capacities for rainwater or well water storage	

Factors to Consider Before Transitioning	Examples	Comments/ Additional Notes
Sanitation	Sustainable recycling of greywater streams for garden irrigation	
	Establishing self-contained blackwater treatment solutions like composting toilets or bio-digesters	
Food Autonomy	Planning diverse crops throughout the seasons for continuous harvests	
	Composting and intensive gardening methods for soil fertility and pest control	
	Regenerative soil care through cover cropping, mulching, and minimal tillage	
	Including resilient livestock and beekeeping for meat, dairy, eggs, and pollination services	
Economic Considerations	Transitioning incrementally by starting with small-scale renewable energy and water systems	
	Leveraging project economics through bulk purchases, group installations, or DIY construction	

Factors to Consider Before Transitioning	Examples	Comments/ Additional Notes
Economic Considerations	Budgeting for upfront transition costs like solar panels, water tanks, or composting toilets	
	Managing operating costs efficiently by optimizing energy usage, water consumption, and waste management	
Mental Transition	Cultivating a growth mindset by embracing challenges as learning opportunities	
	Embracing adaptability and discomfort as part of the journey to self-reliance	
	Building intentional communities for support, knowledge sharing, and collaboration	
	Fostering collaboration and knowledge-sharing through workshops, skill exchanges, or online forums	
Mental Transition	Activating natural learner instincts by seeking new skills, knowledge, and experiences	

This is a simple layout to get the ball rolling and your mind ticking. Feel free to expand on it accordingly to suit your own unique needs.

Embracing the off-grid lifestyle means shattering the illusions of comfort and convenience that society has spoon-fed us for generations. It means rewiring your brain to embrace adaptability, resilience, and an insatiable hunger for growth. Just remember, you're part of a vibrant community of skill-sharers and wisdom-keepers who are all united in their quest for a more sustainable, fulfilling way of life.

CHAPTER 3
WATER COLLECTION AND PURIFICATION PROJECTS

We'd be nowhere without addressing one of life's most primordial necessities—water.

Whether dark storm clouds gather or punishing droughts play out, mastering the art of collecting and purifying this liquid gold is the difference between thriving independently and suffering a slow, bleak decline.

This chapter is all about achieving water autonomy within your off-grid homestead. No more reckless drinking from municipal reservoirs supplied by doubtful stream sources. You're about to equip yourself with practical knowledge for securely establishing a self-replenishing, freshwater supply.

Building Rainwater Harvest Systems

When transitioning towards a fully self-sufficient water supply, few practices prove as empowering and cost-effective as implementing rainwater harvesting systems straight from Mother Nature herself.

Capturing and storing rainwater will provide you with a secure renewable source of freshwater perfectly suited for residential, agricultural, and emergency use. This provides more self-reliance benefits than a McDonald's Happy Meal has plastic toys!

- Rain Barrel Setups

For those just starting their self-sufficiency journey, modest rain barrel setups offer an accessible entry point that's not too hefty on the wallet. With just a couple of 50-gallon plastic drums, some basic PVC piping components, and a raised stand or platform, you can begin collecting runoff from your rooftops during storms. The process is pretty straightforward:

a. Materials: You'll need:

- Rain barrels (the number depends on your water needs and space)

  » Downspout diverters/rain chains

  » Gutter extensions (if necessary)

  » A saw and drill

  » Hose clamps

  » A mesh screen/cover for the barrel inlet

b. Location: Select a spot close to your home's downspouts (the vertical pipes draining rain off the roof), enabling the rain barrels to collect water efficiently. Level the area to ensure it's stable.

c. Prep downspouts: Cut the downspouts where you want to install the diverters. Follow the manufacturer's instructions for installing diverters to direct water into the top inlets of the raised barrels.

d. Barrel installation: Place the rain barrels on stable platforms or blocks, elevating them slightly. This allows gravity to assist with the water flow. Simply connect the downspout diverters to the top inlets of the barrels securely.

e. Overflow system: If you're using multiple barrels, set up overflow mechanisms between them. This can be as simple as connecting a hose between barrels or installing overflow valves. As the barrels fill up, any additional overflow can spill into subsequent barrels arranged in a series.

f. Secure your system: Use hose clamps or secure fittings to tighten and leak-proof connections. Secure a mesh screen/cover over the barrel inlet to keep debris and insects out.

g. Test: Observe how the rainwater flows into the barrels, checking for any leaks or issues with the diversion system.

h. Use and maintain: Use the collected rainwater for non-potable purposes like garden irrigation, cleaning, or laundry. With proper treatment and purification, this rain barrel water can even be used for human consumption. Periodically clean the barrels and screens to prevent blockages and ensure water quality.

i. Don't think small: Those humble pioneering barrel installations are just the beginning of endless opportunities for smart expansion as your rainwater harvesting skills and experience grow over time. You can install things like

» Rain chains (decorative chains guiding roof runoff into barrels)
» Leaf diverters (screens blocking debris)
» First-flush modules (which channel away the initial contaminated runoff)

All of these optimize water quality and make capturing rain even easier. You could even get more advanced by interconnecting your series of rain barrels through shallowly-buried, rigid pipes. This trenched conveyance network lets you control the distribution of overflow into larger underground cistern tanks for storage.

- Larger Collection Setups

For all the truly ambitious self-reliant visionaries on larger rural properties, using large-scale catchment pavilions and pond designs significantly boosts rainwater harvesting capabilities compared to just using barrels. The following are a couple of considerations:

- Devoting significant square footage to smoothly sloped, gently angled collection surfaces like reinforced concrete pads maximize the ability to rapidly receive inflows from each rainy day. Depending on various factors like annual rainfall in your region, the efficiency of the catchment system, and the storage capacity, a properly designed 1,000-square-foot roof-like catchment area can consistently yield over 60,000 gallons of rainwater per year in temperate climates. A serious game-changer for self-sufficiency!

  » Implementing large-scale containment components like in-ground cistern batteries or surface retention ponds becomes critical for wrangling and storing those massive influxes once collected.

  » For larger-scale operations, it's a good idea to consult local water experts or resources specific to your area to get accurate, detailed info on your rainwater harvesting potential.

  » Storage tanks made from durable, UV-resistant materials like HDPE plastic, fiberglass, or ferro-cement provide safe, long-lasting water storage for residential use.

  » You'll also want distribution pipes feeding from the cisterns, potentially pressurized lines for easy access, plus either active filtration components or passive sand filters for purifying stored water before you ultimately use it. Build it right and your entire harvested supply stays in tip-top shape.

No matter which rainwater harvesting approach or scale you choose, there are some universal best practices to follow for optimal results:

- Always carefully position your catchment surfaces and inlets to avoid potential contamination sources like animal feces, fertilizer-laden agricultural runoff, or industrial pollutants and emissions drifting through.

- Use only food-grade, non-toxic containment materials approved for potable water storage, especially anything slated for residential consumption post-treatment.

- Be sure to integrate robust purification protocols, methods like ozonation, UV sterilization, or chemical treatment regimens with something like chlorine, to ensure your captured rainwater achieves reliable consumption standards before any drinking, cooking, or bathing use.

From there, the design possibilities are wide open to express your creativity with rainwater catchment. For example, you could landscape contour through earthworks to create stunning underground cisterns or design cascading rainwater features around your homestead for both beauty and functionality.

You could even choose to stack functions by diverting and storing harvested rainwater into designated tanks for aquaculture or irrigation in your food crops. The potential applications are truly open-ended.

Irrespective of preferred methodology or style, adopting some form of residential rainwater harvesting fundamentally empowers your household's self-sufficiency in various ways which include:

- Nurturing a deep appreciation for your local climate patterns and natural hydro-cycles that often get overlooked and taken for granted with urban living.

- Inspiring consciousness around daily consumption rates and rationing practices since you've become directly involved in your water provisioning.

- Developing a practical understanding of gravity flow and distribution mechanics for effectively managing rainwater across your property.

- And ultimately, it sharpens your ability to observe weather patterns, countering the commercialization of weather data for profit by making you more attentive to local weather in a practical way.

Allow me to reiterate: whether you're starting with those modest beginner rain barrel setups or aiming for advanced catchment systems, prioritize capturing rainwater for your off-grid vision.

DIY Water Purification Techniques

Whether you're living the suburban dream or roughing it in the wilderness, another inescapable truth remains: achieving true self-reliance means having backup methods for turning any available water source into clean, safe drinking supplies.

That life-giving H2O we all take for granted can turn into a putrid, disease-ridden nightmare if it's contaminated by bacteria, viruses, chemicals, or other contaminants.

- Let's start with a total classic—boiling. One of the simplest yet most effective purification principles out there! Bringing water to a good rolling boil for 1 full minute is guaranteed to neutralize upwards of 99.9% of bacteria, viruses, and even stubborn protozoan cysts like Giardia through thermal destruction. The biggest perks of boiling are its universal availability (you just need heat) and essentially zero cost, making it ideal for wilderness scenarios or when you're hunkering down at home during emergencies. Just keep in mind that boiling can potentially concentrate any mineral or chemical contaminants present, and obviously, you'll need a way to heat things up.

- For those who want faster, hands-off purification without hovering over a hotplate, look into chemical treatment options using basic household supplies. A classic method is simply adding a few drops of plain, unscented chlorine bleach per gallon of raw water, giving it at least 30 minutes of contact time before use. Chlorine's oxidizing properties are effective at neutralizing microbes for short-term disinfection needs. Using pre-dissolved chlorine dioxide tablets can achieve faster treatment with less chlorine aftertaste. Just remember, you'll need to keep restocking those consumable chlorine supplies.

- For more serious long-term self-reliance, it's hard to beat the resilience and efficiency of gravity-fed filtration systems. Anything from rudimentary sediment traps made by layering sand, gravel, and charcoal together, all the way up to fancy multi-stage ceramic microfilters with killer pathogen removal capabilities, these setups are a solid choice. The passive gravity-flow nature and indefinite reusability of these filter systems make them incredibly attractive for remote homesteading. With some basic plumbing components and tanks, you could even automate systems that separate pre-filtered and finished polished water reserves while maintaining pressure. Building a gravity filtration system for off-grid living involves various steps. Here are the fundamental steps to give you a better idea of the process involved:

a. Materials: You'll need:

» Two food-grade plastic barrels (one for dirty water, one for filtered water)

» Depending on water needs and available space, a common size for rainwater harvesting barrels is around 55 gallons each.

» Filter media (activated carbon, sand, and gravel)

» PVC pipes and fittings

» Water hose/tubing

» Hose clamps

» Silicone sealant

» Drill and drill bits

b. Prep barrels: The two barrels should ideally be placed side by side or close to each other. Start prepping your barrels as follows:

» For the outlet pipe, drill a hole near the bottom of the dirty water barrel.

» To prevent the overflow pipe from overflowing, drill another hole near the top of this barrel.

» Repeat the same process for the clean water barrel, but this time, only drill holes for the inlet and outlet pipes.

c. Install pipes: Next up, is pipe installation.

» For the outlet pipe that connects to the filter system, attach a PVC pipe with a hose clamp to the hole at the bottom of the dirty water barrel.

» To access filtered water, connect another PVC pipe from the clean water barrel's outlet hole to a spigot or hose.

d. Filtration system: Your filtration system is set up inside the dirty water barrel.

» Layer the filter media (activated carbon, sand, and gravel) in the following order:

Gravel at the bottom

Followed by sand

Activated carbon on top.

• Securely separate each layer with a screen/mesh to prevent mixing.

e. Connect the system: Now, for pulling it all together:

» From the dirty water barrel, connect the inlet pipe to the top of the filter media in the filtration barrel.

» Create a gravity-fed flow from the dirty water barrel through the filter media and into the clean water barrel by making use of PVC pipes and fittings.

» Seal and test: Seal all pipe connections and drilled holes with silicone sealant to prevent leaks.

» After filling the dirty water barrel with water, analyze the flow through the filtration system into the clean water barrel.

» Regularly test to maintain the filtered water quality and adjust the filtration media as needed for optimal results.

• Want to absolutely annihilate any potential pathogens? Concentrated solar thermal purification techniques like the SODIS (Solar Disinfection) method or full-scale solar still distillation units are pretty much the off-grid holy grail. SODIS harnesses the power of sunlight's UV-A rays and heat to disinfect contaminated water just by leaving it contained in basic plastic bottles or bags out in direct sunlight for 6+ hours. You can then take that pre-treated SODIS water and run it through small solar still units that use a greenhouse heat-trapping design to fully evaporate and re-condense it into surgical-grade distilled water.

Don't be afraid to get creative by combining multiple methods into redundant purification conveyor belts. For example, you could do SODIS pre-treatment feeding into multi-stage sediment and ceramic filter banks. Or, have rooftop rainwater first captured, dosed with chlorine, charcoal filtered, and then sent through a solar thermal disinfection blast before reaching sufficient storage vessels. With some clever planning for overlap between methods, you can guarantee endless clean water security for your homestead.

Whichever specific DIY purification options you prioritize based on skills and resources, every technique fortifies valuable abilities into your off-grid toolkit. For instance:

- Properly dosing chlorine helps build chemistry smarts and dilution skills.

- Constructing and maintaining filters instills core plumbing and mechanical handyperson skills.

- Solar distillers impart fundamentals of thermodynamics, fabrication, and big-picture systems thinking.

Using smart methods to cleanse contaminated resources and return them to their original purity empowers you to tap into a near-unlimited resource. So, take time exploring and mastering as many of these DIY purification techniques as you can.

Creating Emergency Water Reserves

Every savvy homesteader worth their salt understands the vital importance of squirreling away dedicated emergency water reserves. When that inevitable "oh dear" curveball gets thrown your way—be it from infrastructure collapse, natural disasters, or whatever the world conjures up—having easily accessible hydration supplies on hand could literally mean the difference between life and drying up like an old prune.

From picking the perfect long-term storage containers to treatment regimens that ensure years of viable shelf life, you'll need to master the fundamentals of drought-proofing your territory against running dry.

There are a bunch of key considerations to cover such as:

- Sourcing Food-Grade Storage Vessels

The containers housing your liquid gold matter immensely.

- Prioritize certified food-safe plastics like high-density polyethylene (HDPE) that won't leach toxic chemicals into your reserves.

  » Stainless steel, fiberglass, and enameled opaque drums provide robust long-lasting options too when properly cared for and regularly inspected.

  » Avoid volatile metals prone to rusting like plain untreated steel.

- ### *Calculating Reserve Capacities*

Determine your household's realistic water consumption rates so you don't accidentally shortchange your supplies.

- For basic unmaintained survival needs like drinking and hygiene, budget around 1 gallon per person daily as an absolute minimum. However, rather realistically plan for 2+ gallons each to accommodate cooking, cleaning, and maybe some gardening.

  » If you're planning to hunker down long-term, strive to stockpile at least two weeks' worth of reserves that are ready to deploy at a moment's notice.

- ### *Sizing Storage Assets*

Based on your customized consumption calculations, start mapping out appropriate container sizes and overall volume requirements.

- The classic 55-gallon HDPE "water brick" containers provide a nice sturdy supply with Tetris-like stacking density.

  » If space allows, larger 150-300 gallon HDPE totes may maximize stockpiling efficiency too.

> » Look for containers with built-in level transparency to easily visually monitor liquid-level gauges.

- *Prepping and Treatment*

Even brand-new containers require full sanitizing rinses to remove any potential manufacturing residues or contaminants before being used.

- Simple food-grade household bleach provides an affordable, effective sterilization method at just 8 drops per gallon with 30 minutes of contact time before a good fresh water flush.

  > » For the hardcore survivalist preppers out there, glucose supplements could be added to stimulate anaerobic bacterial spore growth for extra shelf-life longevity.

- *Rotation and FIFO Practices*

Just like rationing any precious resource, you have to employ a strict First-In-First-Out (FIFO) inventory rotation system to ensure optimal water quality and longevity.

- Prioritize using up the oldest stockpiled reserves first before cracking into fresher replacements. It's rather logical.

  > » Set calendar reminders to periodically cycle through supplies, replenish anything already consumed, and make mindful incorporation of annual stockpile maintenance a part of your seasonal homesteading routines.

- *Climate-Controlled Storage*

This one's crucial: avoid exposing your reserves to extreme temperatures, as this can lead to bacterial growth and plastic deterioration.

- Find a nice cool, temperature-stable environment like your basement's deepest recesses or a properly-insulated root cellar for storing your H2O caches.

  > » Closets, laundry rooms, and other interior spaces can work too, depending

on temperature.

- *Accessibility & Security*

Keep the proximity and discretion of your reserves from common household areas in mind.

- Disperse supplies throughout various nooks and crannies enables ease of access while mitigating the risk of dreaded single-point failures. You know, nasty surprises like earthquakes, floods, or fires?

  » Get creative about obscuring storage from potential intruder eyes through cunningly hidden panels, disguised shelving or even buried underground rain-safe bunkers.

Centralizing your water reserves and strategically stashing them throughout your home can ensure you have enough water to see you through man-made or nature-induced time. So study up on these critical storage practices and start making iterative implementations to stay hydrated.

Sustainable Water Usage Practices

Securing long-term control over water isn't just about sourcing and stockpiling it. Using it wisely every day and conserving it is equally vital to reducing reliance on external sources.

Adopting some brazenly wasteful, carefree attitude around managing your hard-earned water reserves is just straight-up self-defeating. Even with redundant rainwater harvesting, well drawdown, and atmospheric capture capabilities already in place, you need to learn how to conserve your water to make sure you're not using and abusing your resources. Here are some strategies on how.

Mindful consumption tracking: One of the smartest habits is simply instituting daily monitoring of your residential withdrawal rates alongside the amount of rainwater replenishing your reserves.

Efficiency upgrades & retrofits: Do a systematic audit to optimize water usage in your home by fixing leaks in appliances and fixtures, installing aerators to reduce flow rates, and identifying opportunities for pressure reduction. Prioritize sensible upgrades like low-flow showerheads and faucet aerators, on-demand tankless water heaters, and front-loading high-efficiency washers.

Greywater reuses: Stop mindlessly disposing of those "greywater" streams from laundry, dishwashing, and showers. Get creative and reroute those already-lightly-used nutrient-dense flows out into surrounding garden guilds and edible landscapes instead of just letting them drain away as waste.

Smart landscaping: Switch out water-thirsty grass and decorative plants with xeriscaping, a landscaping method that conserves water using drought-resistant plants and efficient irrigation, tailored to your area's climate. Implement contoured swales, berms, and micro-basins to guide rainwater toward plantable areas, promoting natural water retention and efficient use.

Regenerative growing: It's not enough to just conserve and avoid waste. For true ecological hydro-integration, you'll want to explore regenerative cultivation principles that actively harmonize your gardening and growing efforts into holistic natural moisture cycles. Techniques like hugelkultur (a gardening technique that involves creating raised beds filled with decomposing wood, organic matter, and soil), wicking bed geometries, and strategic soil amendments like moisture-retentive biochar, blend human systems with natural water flows.

Water storage integration: Don't view those rain barrel catches, cistern reserves, and emergency potable stockpiles as static utilities separate from your landscape. Thoughtfully situate these water storage assets throughout growing zones so they can anchor gravity-fed dispersion, drip irrigation, and contouring strategies.

In the end, true self-reliant water autonomy extends far beyond just sourcing supplies and then thoughtlessly consuming rations. It's about transcending the mindset of taking from nature and seeing water's role in sustaining life on land as abundant and regenerative.

CHAPTER 4
FOOD PRODUCTION: GARDENING AND LIVESTOCK MANAGEMENT

Playing pioneer homesteader makeover with rainwater harvesting and off-grid power systems will only get you so far on this self-reliance journey. At the end of the day, even the most high-tech hydroponic greenhouse or tricked-out solar microgrid won't mean diddly-squat if you can't actually put nourishing, life-sustaining food on the table.

Off-grid living means nurturing a green thumb and blending your living spaces with productive natural areas in harmony. This integration allows you to enjoy a self-sufficient diet full of quality, integrity, and soul.

Permaculture: Creating Self-Sustaining Gardens

Traditional row gardening might keep you busy, but it does not equate to true self-sufficiency.

Then what's the true secret to cultivating nourishment for your household while maintaining your sanity? Permaculture—a smart way to use plants that can handle dry conditions, saving water and reducing the need for constant care. In a nutshell, permaculture is all about intelligently designing gardens, landscapes, and even entire human environments that harmoniously integrate and cooperate with the natural energies, patterns, and cycles of the local ecosystem. This approach goes against conventional farming that dominates and alters nature to fit human preferences.

Instead, permaculture is all about establishing cultivated systems that simply build upon, enhance, and collaborate with what already works based on the specific climatic, and soil conditions, and inhabitant species present within a given bioregion. You learn from and copy sustainable ideas, then add the food you want to grow naturally instead of forcing them into the environment. These design principles are applied repeatedly across all scales, from a single urban gardening bed to integrated master plans.

- Conscientious Patterning

Everything in nature manifests through geometry, the dynamics of waves, meanders, branches, convection cells, you name it. When applied creatively to garden layouts and agricultural landscaping, strategic patterning can boost the benefits and create helpful connections between elements. For example:

- Carefully positioning plantings using zonal methods, relative location analysis, and sector mapping optimizes outputs and resource allocation like water, nutrients, and labor.

  » Planning curvilinear paths, terracing, and graded slopes enhance hydration and reduce erosion.

  » Thoughtful geometries like spirals or fractals amplify ecological production.

- ## *Comprehensive Resource Stacking*

Rather than conventional mono-cropping, which extracts single yields through mining the soil, permaculture cultivates diverse polycultures that use vertical growing space efficiently. This method maximizes edible yield in every available space as follows:

- Physical stacking across canopied layers where the tallest tree crops provide shade and passively prune lower branches of smaller fruit trees and shrubs below.

  » Herbaceous perennials and groundcovers carpet and cover empty open areas.

  » Underground spaces are used for crops like tubers and mushrooms.

- ## *Dynamic Self-Regulation*

The ultimate goal of permaculture is for self-sustaining systems that recycle resources internally, needing minimal human input. For example:

- Maybe your banana circle combines nutrient-depositing chickens, outer areas for biomass, and a vermicomposting center for herbs and vines. This could yield edible crops each season in a mutually beneficial way.

  » A food forest could have its own soil aerators, fungal decomposers, and pollinator habitats for healthy soil and plant growth.

Cultivated permaculture communities mimic mature natural ecosystems by gradually becoming self-sufficient in water, nutrients, soil fertility, and resilience. Continuous observation and strategic adjustments enhance their self-sustainability over time.

Of course, I must include some essential tips for first-time permaculture enthusiasts who are just starting out:

Thoroughly assess your specific site: Before diving into stone-laying or tree-planting, you must conduct a site analysis covering aspects like:

- »   a. Climate patterns
- »   b. Seasonal solar mapping
- »   c. Hydration flowlines
- »   d. Existing vegetation
- »   e. Soil composition
- »   f. Water sources and drainage
- »   g. Microclimates
- »   h. Slope and topography
- »   i. Wildlife habitat
- »   l. Access and infrastructure
- »   m. Legal and regulatory considerations

Carefully document this observational data as a reference library to create ecologically aligned designs.

Start with annual vegetables: Rome wasn't built in a day, and neither are full-scale permaculture patches. Good things take time. Start small with basic vegetable beds, planting both transplants and directly seeded crops. Here's a simple guide on how to build soil fertility and embrace composting:

a. Location: Choose a suitable location for your compost pile or bin that is well-drained and placed in a partially shaded area to prevent drying.

b. Materials: Collect a mix of "greens" and "browns," aiming for a balanced ratio of roughly 2:1 or 3:1.

- »   Greens: Nitrogen-rich materials like kitchen scraps, grass clippings, and coffee grounds.
- »   Browns: Carbon-rich materials like dried leaves, straw, and shredded paper.

c. Build the base: Cover the bottom of your composting area with a layer of browns to provide aeration and drainage. This will help prevent the pile from becoming too compacted.

d. Add greens and browns: Add greens and browns as they become available, alternating layers. Be sure to shred larger materials for speedier decomposition and keep the pile moist, but not soggy.

e. Shovel the pile: Regularly turn and shovel the compost pile for aeration, decomposition, and temperature regulation.

f. Monitor: Regularly check the moisture level. It should feel damp and prevent the pile from becoming waterlogged. The internal temperature of the pile should ideally clock in between 120-160°F (49-71°C) for optimal decomposition.

g. Cover: Consider covering the compost pile with a tarp or lid so that it retains moisture and heat, especially in dry or cold climates. This will also help accelerate composting.

h. Patience: Depending on various factors like the size, materials used, and environmental conditions, composting can take weeks to months to properly decompose. When it's dark, crumbly, and earthy-smelling, with no recognizable materials, it's ready to use.

i. Use: Once the compost is ready, mix it into your garden soil, or potting mixes, or use it as a top dressing for plants. Never use immature compost that still has recognizable materials, it can rob plants of nitrogen during decomposition.

Establish hardscape infrastructures: Once you've found the best areas you can start setting up the main farming layout and building permanent features like ditches, leveled areas, water systems, paths, and planting spaces before adding perennial plants.

Plant with succession in mind: Plan your garden's growth from start to finish before getting your hands dirty. Start with components like nitrogen-fixers and overstory trees, then gradually add layers to create a thriving ecosystem. A self-sustaining forest can't be rushed.

Nurture intelligent poly-guilds: In nature, every organism works together in beneficial groups. Mimic this idea to your crops by planting them in communities that support each other, providing resources, airflow, and protection.

Gently ease into self-reliance: Start using low-intervention techniques such as sheet mulching, natural fertilization with chop-and-drop, and integrated pest control.

At the end of the day, these permaculture methods go way beyond just hobbyist gardening. They create resilient systems that not only produce food but also restore the environment.

Raising Small Livestock for Food Security

What's not to love about tending to permaculture gardens overflowing with edible abundance? It's like having your own personal farmers' market right in your backyard!

But there's a hard truth—plant calories alone don't make a wholesome, nutritionally complete diet. Sure, loading up on vitamin-packed produce and heritage grains is a solid foundation for food independence, but if you want to reach that fully self-sufficient, nutrient-rich holy grail, you've got to integrate some resilient protein sources into the mix.

This is where small-scale livestock operations step in. Not only do they provide fresh animal proteins like eggs, meat, and dairy products on demand, but you're also getting some valuable "byproducts" that can be repurposed as self-fertilizing soil amendments, turbocharging your edible landscapes. Talk about a win-win!

Let's start with some simple livestock selection factors to consider:

• Space and Housing Requirements

Even urban micro-homesteaders dwelling on petite plots can incorporate poultry flocks or meat rabbits. Slightly larger suburban settings can expand the menu to incorporate miniature dairy goats or micro herds of swine.

Just be sure to thoroughly research each species' unique spatial and infrastructure needs before welcoming any farmyard guests. Coops, hutches, runs, and loafing areas all demand accommodating footprints. That said, vertically incorporated housing solutions help maximize tight spaces efficiently.

• Local Climate Considerations

While chickens are adaptable, factors like seasonal temperature extremes, rainfall patterns, and elevation variables definitely influence ideal breed suitability.

For example, dense-feathered birds like Chanteclers handle frosty northern winters far better than skimpier egg-layers like Leghorns or Anconas. Spanish goats with their sleek coats cope well with scorching summer conditions compared to heavy-wooled Angoras or Toggenburgs. So assess your specific microclimate before committing to any specific breeds or species.

• Feed Availability and Diet Demands

Nothing goes against self-sufficiency advantages of small livestock than relying on expensive, bagged feeds.

The whole point is to use resources on-site efficiently without relying too much on external inputs. Take stock of your existing acreage to see if your land can provide extra food for your animals, such as hay, plants from cover crops, leftovers from processing food, and plants regrowing from gardens. Dual-purpose birds and bovids are excellent nutrient recyclers and help fertilize soils naturally through grazing rotations!

• Byproduct Applications and Value-Adding

Speaking of self-fertilizing assets, don't overlook the incredible regenerative gold mine within nutritious animal manure, bedding materials, and even culled offal byproducts.

Sure, goat droppings and poultry litter generate potent fertilizer to supercharge your veggie beds. But, do yourself one better by designing integrated farming systems that turn your waste into valuable opportunities. For instance, funnel manures into methane biodigesters, producing livestock barn heat, and cooking fuel. Or, turn used bedding into food for animals through vermiculture bins to create protein-rich supplements.

• General Suitability Personalities

Not all livestock candidates work well with people equally, so consider their compatibility with your lifestyle.

For example, those social dairy goats with their charming personalities are ideal companions and coveted milk producers for families. Larger bovines like cattle or water buffalo might demand too much daily commitment. Docile laying hens also make ultra-low-maintenance family friends, while free-ranging heritage turkeys gobble up pest insects, clearing gardens.

With due diligence in accounting for all those core selection factors, there are also the logistical livestock topics like:

• Housing Design and Construction

When setting up your backyard farm, focus on essential aspects like cleanliness, safety, and nutrition.

For chickens, build predator-proof coops with roosting bars and nesting boxes before expanding. Goats and sheep require sturdy fences deterring escapees, along with sheltered barns for resting and chewing cud. Always provide reliable water sources, natural ventilation, insulation against the weather, and sufficient dry bedding for all animals.

• Feeding and Fodder Practices

Although convenient pellet rations are hard to avoid occasionally, whenever possible, prioritize getting nutrition from on-site fodder.

Home-scale brewers can supplement spent grains into poultry diets while fermenting food scraps into probiotic liquids that are beneficial for ruminants' health. On-pasture rotational grazing is a powerful way to rejuvenate soils and cut down on feed costs.

- Preventative Healthcare and Grooming

As with any living being, keeping small livestock healthy and productive requires preventative health care routines, following the "prevention is better than cure" approach.

Beyond barn cleanliness, basic practices like regular deworming, inspecting for injuries, grooming and hoof trimming, and integrated pest management, help prevent costly disease outbreaks. Small farmers should buddy up with local veterinarians or healers and herbalists, assisting with interventional treatments when needed. But, keep in mind that resilient livestock breeds matched to the environment rarely require constant pharmaceutical crutches.

- Ethical Practices and Processing

Should your food security plan include sustainably incorporating animal proteins through humane culling, prepare to confront the realities of life-cycle philosophies head-on with honesty.

Learn low-stress handling techniques to minimize stress responses. Study anatomical butchery enabling "root-to-leaf" waste management like rendering fats and tanning hides. Commit to offering humane dispatching without unnecessary suffering, showing respect for the food sources.

Plan well by choosing the right animals, planning their food, setting up systems, and being ready for chores. Nature invites you to co-create abundance, so accept this empowering mantle and welcome small livestock into your off-grid life.

Essential Food Preservation Techniques

All this talk about cultivating lush permaculture food forests and integrating hardy livestock into regenerative cycles is great and all, but it's only half the food freedom equation.

Because at the end of the day, even the most abundant harvests and "egg-tastic" poultry outputs reach a peak before trailing off into scarce winter slumps. And if you haven't mastered the art of preservation, all those glorious summer harvests will soon be nothing more than a compost pile of unfulfilled dreams.

That's exactly why developing an arsenal of time-honored food preservation techniques is a non-negotiable skill set for any household striving towards full year-round self-reliance. Why let all that hard-earned seasonal bounty go to waste?

Everything from moisture-removing dehydration methods to anaerobic fermentation that concentrates flavors into flavor-rich superfoods can help preserve your foods through cooler months.

- Water Bath Canning

This classic preservation approach uses pressurized steam heating to safely acidify food contents within airtight sealed jars. Perfect for salsas, jams, pickles, and certain tomato products.

Canning represents the premium solution for extending shelf life by 1-5 years with minimal quality degradation. Just mind your protocols carefully. Issues like improper head spacing, inadequate thermal processing, or sloppy sanitation can easily spoil canned goods. When done right, those coveted pantry fruits and veggies gain extended fridge-free storage capacities. Just take a look at how simple canning is:

Equipment: Water bath canner (a large pot with a lid and a rack that holds the jars)

» Canning jars, lids, and bands

» Jar lifter/tongs

» Canning funnel

» Lid lifter or magnet wand

» Clean towels and kitchen cloths

Recipe: Select a recipe suitable for canning (jams, jellies, pickles, or sauces).

» Gather all the ingredients and follow the recipe prep instructions.

Sterilize and Prep: Thoroughly wash the jars, lids, and bands in hot, soapy water.

» Rinse it all properly and place the jars in the canner filled with water, fully submerging them.

» Bring the water to a boil, sterilizing the jars for a minimum of at least 10 minutes.

» To soften the sealing compound, keep the lids and bands in a separate pot of simmering water.

Fill: Use your canning funnel and carefully fill each sterilized jar with your prepped recipe.

» Remember to leave the recommended headspace specified in your recipe which is usually between 1/4 to 1/2 inch.

» Run a non-metallic utensil along the sides of the jar to remove any air bubbles.

Seal: Remove any residue from your jars, wiping the rims with a clean, damp cloth.

» Place the lids on the jars using a lid lifter.

» Screw on the bands until they are snug but not overly tight.

Process: Place the filled and sealed jars back into the canner, using the jar lifter. Be sure that the jars do not touch each other or the sides of the canner.

» The jars must be covered by at least 1 to 2 inches of water.

» Bring the water back to a boil, starting the processing time according to your recipe and altitude.

» Maintain a steady boil throughout.

Cool: Once the processing time is over, turn off the heat and carefully remove the jars.

» For air circulation, place and space the jars on a towel-lined surface or cooling rack.

» Let the jars cool for 12 to 24 hours.

Check Seals: Press down on the center of each lid to test the seals. It's sealed if the lid doesn't pop or flex.

» Label and date the sealed jars.

» Store in a cool, dark place.

» Any unsealed jars should be refrigerated and consumed timely.

• *Dehydrating and Drying*

For shelf-stable lightweight portability, dehydration reigns supreme at rapidly removing moisture content from foods.

Modern advanced electric appliances simplify the process down to foolproof operations, but the lower-tech solutions, like solar dehydrators or wood-stoked evaporators, integrate beautifully with off-grid homesteads. Jerky enthusiasts especially value the transformative power of drying, turning excess meat into highly concentrated protein snacks that last for months. Similarly, herbs become potent teas, fruits transform into chewy rolls, and vegetables are reborn as crispy chips. Endless dried possibilities abound here!

• Root Cellaring/Cold Storage

If you've already put in the work to excavate proper root cellars or larders beneath your homestead's thermal mass, using their naturally cool environments for temporary food storage is a practical choice.

Root crops like potatoes, carrots, and beets keep splendidly under these conditions for months on end. The same applies to cured onions, shallots, and garlic bulbs. Some folks even experiment with very low-humidity layouts for stockpiling winter squashes, cabbages, and apples.

- Freezing and Cryogenics

On the opposite end of the temperature spectrum, freezing rapidly locks in food flavors by immobilizing moisture content into solid states.

Ultra-low subzero temps induce something called cryogenic preservation, enabling cellular structures to remain intact for years. While homestead freezers demand power sources and uninterrupted upkeep, their convenience for meal-prepping delicious make-ahead dishes simply can't be denied for busy families. Just monitor those seals, blanch veggies first, and embrace those zippered vacuum-sealing contraptions.

- Lacto-Fermentation and Culturing

Here's where things get delightfully funky and scientifically fascinating. Lacto-fermentation helps good bacteria grow, turning sugars into tasty acids that keep food fresh without chemicals.

Sauerkraut, kimchi, kvass, and kosher dills all celebrate those microbial transformations that build flavor while extending shelf life indefinitely in anaerobic environments. Fermented beverages like kombucha make for refreshing probiotic potions too.

- Smoking and Curing

This skill takes some seasoned experience to master properly. Leveraging smokehouses to infuse meats, fish, and other proteins with antimicrobial compounds, like nitrites, through extended drying and curing practices elevates off-grid provisions into artisanal bites.

While fire management and temperature monitoring require diligence, mastering these techniques produces downright tasty treats like candied salmon jerky, smoked barrel-aged cheeses, and charcuterie boards your ancestors would drool over.

Master these traditional preservation practices and you'll safeguard your household's sustenance channels.

Seed Saving and Crop Rotation

At this point, you've absorbed a boatload of homesteading wisdom spanning permaculture design, small livestock farming, and even culinary preservation arts.

But before those self-reliance egos start inflating too big for the barnyard, allow me to humble you real quick with a stark realization: even the most prolific off-grid gardens and farm operations are just temporary blips on the radar without intentional agricultural cycle management. Sustainable long-term success requires handling big-picture fundamentals like seed independence and strategic crop rotations. Skipping out on these seemingly small but crucial cultivation maxims is a surefire way to gradually degrade your hard-won growing systems into barren wastelands faster than a plague of locusts.

Let's start with unpacking the all-important practice of reservoir seed saving and stewardship. At the most basic level, learning how to collect, cure, and properly store viable seed reserves from your best plant outputs ensures self-replenishing crop genetics year-over-year without codependency on industrial suppliers. This skill set involves understanding basic botany concepts such as pollination and seed formation, as well as carefully selecting parent plants with desired genetic traits for successful propagation. Isolating, bagging, drying, winnowing, and storing those precious seed reservoirs season after season breeds wholesome independence.

You need to use local seeds that are used to the weather and diseases in your area to grow crops successfully without relying on outside resources. This is where the practice of seed saving brilliantly connects to the complementary philosophy of strategic crop rotation planning.

At the most basic level, intelligent crop rotation helps maximize yields by intentionally alternating different plant families across your annual growing beds in multi-season successions. This doesn't just avoid depleting soil reserves of specific nutrients over time; it actively builds long-term soil health through symbiotic relationships like nitrogen fixation from legumes. It also helps mitigate pest and disease exposure by confusing their life cycles.

Crop rotations are greatly enhanced by following guidelines that avoid planting the same crop year after year in the same place. Instead, you'll be integrating biodiversity through interplanting polycultures of complementary companion crops that permaculture folks call "stacking functions."

Think edible, low-growing vegetation planted under vine crops that are supported by trellises. These vines are under taller plants, like corn, that need a lot of nutrients. Climbing legumes do well by growing up corn stalks, helping to put nitrogen back into the soil. From there, your rotation game plan evolves into dividing farmable areas into dedicated sections for things like cereals, grains, legumes, nightshades, leafy greens, and so on.

Beds get productivity boosts through periods of natural cover cropping and even livestock rotations interspersed with rejuvenating fallow sod-recharging seasons. It's like agricultural choreography!

And the payoffs for diligently sticking to these time-tested rotation scripts?

- Pests naturally dissipate as their specific hosts continuously move around.

- Soil nutrient levels experience consistent organic boosts through diversity.

- Weeds struggle to establish footholds amidst the constantly shifting botanical diversity.

- Yields increase and there's more abundance in the environment over time.

By saving seeds that suit your area and planning how crops rotate, you can create a self-sustaining farm that mirrors nature's healthy balance. The result? Continual bountiful harvests, fulfilling needs season after season.

This level of food system mastery takes diligent multi-year planning and record-keeping well beyond short-term gratification. You'll need to keep refining your observations as you go through different growing seasons, learning and adjusting along the way. It sounds like piling more work onto already complex homesteading duties, but I really encourage you to embrace these agricultural cycle practices from the very start.

CHAPTER 5
RENEWABLE ENERGY SOLUTIONS FOR YOUR HOME

Self-sufficiency doesn't stop at nurturing your off-grid oasis.

You need energy! We can't stay stuck as powerless debtors to big utility companies that rely on harmful fossil fuels. Breaking away from utilities requires some initial investment of time and resources, but it quickly leads to real self-sufficiency. It also brings your household's consumption rhythms back into harmony with nature's eternal generation cycles that our industrial age so arrogantly disregards and disrupts.

DIY Solar Panel Installation and Maintenance

Let's address the sun-harnessing elephant in the room; residential solar panel systems! What better place to start reclaiming your household's energetic autonomy than by tapping into the prime power source blazing across the sky?

Perhaps you're thinking, "Solar seems awesome in theory, but isn't installation super technical and expensive for us common folk?" Absolutely fair, but, prepare to have those anxieties silenced.

Considerations

- Site Evaluation

Before parting from your hard-earned cash, you'll need to evaluate your site's solar potential. Key factors include:

- Solar exposure: You'll want an area that receives direct sunlight for most of the day, free from shading by trees, buildings, mountains, or any other potential obstructions.

  » Roof orientation: South-facing roofs in the northern hemisphere receive optimal sun exposure.

  » Roof condition: The roof must be in good shape and have a 25+ year lifespan remaining.

  » Local climate: Consider temperature extremes, wind conditions, and weather events like hail storms.

  » Future expansion: Consider scalability options and available space to accommodate increased energy needs over time.

  » A quick tip: An online solar calculator can estimate your site's solar production potential based on location data.

- *System Sizing*

Next, you'll need to determine your electrical needs to size the solar system properly. This involves:

Calculating electrical loads: Determine the energy consumption of your home's appliances, heating/cooling systems, lighting, and other electrical devices.

Estimating future loads: Anticipate future energy demands, like electric vehicle charging (EV charging), to avoid under-sizing the solar system. Also, consider potential additions or upgrades to your household that may increase electricity usage.

Energy efficiency upgrades: Implement energy-efficient upgrades and practices to reduce the overall electricity demand such as using energy-efficient appliances, LED lighting, and improving insulation and sealing to minimize heating and cooling losses.

Panel capacity: Determine the number of solar panels required based on your calculated electrical loads and energy efficiency improvements. Most homes typically need between 10 to 30 solar panels to meet the full demand.

• *Purchasing Equipment*

With sizing complete, you can select and purchase system components such as:

Solar panels: When it comes to choosing solar panels, look at efficiency, degradation rates (how much output decreases over time), and temperature coefficients (performance in different temperatures).

Inverters: Select inverters that can efficiently convert DC (direct current) power generated by solar panels into AC (alternating current) usable in your home.

Charge controllers: If you're incorporating battery storage, use charge controllers to regulate the charging process, and prevent overcharging, and battery damage.

Batteries: You have two options here; lead-acid or lithium-ion batteries. Keep the battery capacity and compatibility with your overall system design in mind.

Mounting system: Mounting racks designed for the roof or ground should be able to withstand local wind and snow loads. Here's a solid example of a panel ground mount:

Materials: Calculate the dimensions and preferred design of your solar panel rack according to your space and solar panel configurations.

You will need:

a.   Galvanized steel pipes

b.   Modular connectors (T connectors and pivot tees)

c   Adjustable joints like the HJ-6

d.   Flat offset pipe clamps

e.   Tools like a hex drill bit set, a pipe cutter, and a t-handle wrench

Cut tubes: With a pipe cutter, cut the steel pipes according to your dimensions. The following lengths are guidelines:

a.   2 pieces: 74 5/8 inches

b.   2 pieces: 59 inches

c.   2 pieces: 36 7/16 inches

d.   1 piece: 42 1/4 inches

Assembly: Connect the tubes using the flat offset pipe clamps and modular connectors.

Ensure correct placement by following the color and letter codes provided with the connectors.

Secure all the joints and bolts, using the hex drill bit set and T handle wrench.

Adjust: Use your HJ-6 adjustable joints and adjust the angle of the solar panel rack to optimize sun exposure.

Install: When your rack is assembled and sturdy, you can mount your solar panels onto the frame.

Brands: Research customer reviews and warranties for more in-depth information about equipment quality and support.

Installation

Solar panel installation involves a couple of key steps:

Racking: Securely mounting rails along your roof's rafters or on ground poles.

- Begin by securely mounting the racking rails that will hold your solar panels in place.

- For roof installations, use strong mounts and bolts designed for your roof type to secure the rails to the rafters that are able to handle windy conditions.

- Proper spacing and orientation angled towards the sun is important.

- For ground mounts, the rails should be anchored to aluminum or steel poles which are embedded into concrete footings or ground screws.

- Frame rails must be perfectly level and square.

Wiring: Connect the cables from the solar arrays on your roof or ground to your home's electrical system.

- Once the racking is in place, the next step is running the electrical wires to connect your solar panels to the inverters and eventually your home's electrical service panel.

- Use UV-rated outdoor photovoltaic cable secured with proper cable clamp spacing.

- When wiring goes through the roof, use specialized flashed roof attachments to maintain a watertight seal.

- Wires should be routed with drip loops to shed moisture away from connections.

- Strictly adhere to electrical code requirements, including installing disconnects and overcurrent protection like fused strings between your solar sub-arrays and inverters.

Grounding: All equipment must be properly grounded as per local electrical codes.

- Proper grounding and bonding are absolutely critical for safety.

- Ensure proper grounding by installing copper grounding electrodes like rods or ground rings near your inverters.

- This protects against electrical faults and lightning strikes by providing a path to safely discharge current directly into the earth before it can become a shock or fire hazard.

Panel mounting: Carefully laying out and bolting panels onto the racking.

- With the racking installed, it's time to carefully lay out and secure your solar panels.

- Lift each panel into position, ensuring the wiring leads neatly pass into the racking channels, and then bolt them down at the proper tilt and orientation.

- Be sure that there's equal spacing between panels to prevent shadowing and hot spots. Use hijinks or quarter-turn fasteners designed for solar modules.

- Double-check all electrical connections are secured and free from ground faults.

Equipment hookup: Interconnecting inverters, charge controllers, batteries, and existing electrical panel.

- Now you'll interconnect and configure the balance of system components like inverters, charge controllers, and batteries (if using a storage system).

- Wire the solar panel leads into the combiner boxes, then connect home runs to inverters and battery banks.

- Integrate the inverter output circuits with your home's electrical service panel using a separate circuit breaker.

- Adhere to manufacturer instructions and electrical codes. Improperly wired systems can create fire hazards and system faults.

Permits: Most areas require approved permits, inspections, and interconnection approval before operating.

- All jurisdictions require permitting and inspections for solar PV installations.

- An electrical permit must be obtained before work starts.

- Once complete, scheduled inspections of your grounding, wiring, and interconnection methods should be analyzed.

- If grid-tied, you'll also need explicit written permission from your utility before interconnecting to avoid accidentally sending power to de-energized grid lines. They'll perform their own inspection and likely require an external disconnect switch.

- DIY installation demands care and safety adherence to electrical codes. For complex jobs, hire a professional.

Overall solar installations are very doable for diligent DIYers, but safety codes absolutely cannot be sidestepped. Complete this process correctly, and you'll be generating clean electricity for decades!

Maintenance

As with all equipment, with solar panels installed, ongoing maintenance is essential:

Weather conditions: Ensure your system can withstand weather extremes and take appropriate preventive measures.

Pest control: Deter birds or other pests from nesting near or on your solar panels with bird spikes or nets for example.

Tree trimming: Keep greenery trimmed that could cast shade on your panels.

Monitoring software: To quickly identify any issues and take prompt action, consider investing in monitoring software or systems.

Professional inspections: Schedule annual or bi-annual professional inspections for more thorough checks on your solar system.

Warranty coverage: Understand what maintenance tasks may void warranties and follow manufacturer guidelines accordingly.

Emergency preparedness: Have a plan B in place for emergencies such as severe weather events or system malfunctions. Know how to shut off power safely if needed.

Battery care: For lead acid, check fluid levels, clean terminals, and replace every 5-7 years.

Recording: Log issues, repairs, and production data to identify trends.

Proper cleaning, inspection, and replacements preserve maximum energy harvest for decades.

Upgrades

Plan for potential solar system expansions or upgrades like

Capacity: Plan ahead for any additional panels or higher efficiency models to meet increased demands over time.

Battery storage: For extra backup powder, consider upgrading to larger-capacity batteries.

Inverter tech: Consider options like micro-inverters or higher-capacity inverters for optimal energy conversion and system performance.

EV charging: Integrate EV charging capabilities into your solar system to maximize your renewable energy use and reduce reliance on grid electricity.

Designing and planning for future adaptability prevents future costly overhauls.

As long as you commit to safety and stick to guidelines, you'll be generating renewable energy for decades!

The Basics of Residential Wind Turbine

Maybe not everyone is quite ready to go full-tilt solar evangelist and baptize themselves in the radiant flows of the sun. All good! For those of you with wide-open spaces and a steady breeze, residential wind turbines might be the perfect way to start your off-grid journey.

We're talking about harnessing the amazing kinetic energy of the wind, and we're not just talking about rinky-dink novelty garden windmills. With some careful planning and the right turbine setup, you can actually generate a significant portion of your home's energy needs just by capturing the power of the wind.

Before we can anoint you as the wind wizard, we'll need to cover some basics. First, you'll need to figure out if your property is even suitable for wind power. You don't want to invest time and money into setting up a turbine only to find out that your location just doesn't get enough wind. If your area does have a solid breeze, you'll need to consider the different types of wind turbines available and which one will work best for your situation.

• Site Scouting 101

The very first order of business involves assessing whether your slice of paradise enjoys enough consistent windy conditions to make investing in a turbine worthwhile. Because as satisfying as it'd feel sticking it to your local utility monopoly, erecting a pricey wind harvester only to face feeble breeze realities would just leave you with a serious case of buyer's remorse.

Some site characteristics indicating solid wind power potential include:

Unobstructed wind paths: Minimal larger buildings, trees, or terrain blocking dominant wind corridors.

Elevation advantages: Properties atop hills/ridgelines exposed to flowing air currents.

Consistent winds: Steady, reasonably strong winds above 7-10 mph on average.

Noise considerations: Consider the noise impact on residential areas or sensitive locations.

Safety factors: Consider safety measures to prevent accidents or hazards related to turbine operation and maintenance.

You can do a basic wind assessment yourself using portable wind meters, but for larger installations, it's a good idea to get a professional site assessment to really map out the wind patterns throughout the year.

• Turbine Architecture 101

Assuming your castle isn't cursed with stagnant air challenges, next comes evaluating what specific turbine architectures best harmonize with your environment's unique conditions. The two primary categories residential wind systems fall into are:

Horizontal-Axis (HAWT)

» These are the classic propeller-style turbines with 2-3 blades that spin parallel to the wind. They're generally more efficient and can capture more energy from the wind, but they're also taller and the spinning blades can be a safety concern.

Vertical-Axis (VAWT)

» These turbines have a helical, eggbeater-like shape and can capture wind from any direction. They're not as efficient as HAWTs, but they can operate closer to the ground and handle turbulent winds better.

Just remember that proper turbine selection depends on your site's wind characteristics, local zoning restrictions, installation budget, and energy offset targets.

• System Sizing and Installation

Once you've picked your turbine and have quantified your household's anticipated electricity consumption, you can choose the right size wind system to meet your needs.

You'll account for baseline loads like appliances plus any future expansions like EV chargers. Turbine manufacturers have calculators that can help you find the perfect size based on your energy needs and wind data. Understanding the size of the rotor, the shape of the blades, and the capacity of the generator is important for extracting the most energy from your wind system. Once you have all the equipment details, focus on installation, which includes:

• Setting up foundations

» Raising the tower
» Connecting wires to controllers and batteries

Concrete pads are typically used to support residential tower designs to handle wind loads. Ensure the wind vane points in the right direction and plan out how cables will run. You might also have to get permits and approvals, especially for bigger turbines that may need cranes for installation.

• Maintenance and Safety

Residential wind systems require periodic maintenance and safety protocols to keep your silky-smooth breeze harvester spinning. Key aspects include:

Routine inspections: Visually scan for blade damage, excessive vibrations, or other irregularities during operation.

Lubrication and cleaning: Keeping rotating components lubed while removing debris prevents premature wear.

Vegetation control: Trimming nearby trees prevents blade strikes and optimizes wind exposure.

Lightning protection: Proper grounding and air terminal bonding prevent equipment damage or personal risks.

Aviation safety: Marking conspicuous towers with warning lights/paint for aircraft awareness. This is important!

Most residential turbine manufacturers provide maintenance checklists and service guidelines covering recommended intervals.

Installing a residential wind turbine is an awesome way to start generating your own renewable energy, but it might not be enough to cover all your electricity needs right away. That's where solar power can come in—by combining wind and solar, you can create a more reliable, well-rounded system that can keep your lights on even when the wind isn't blowing.

Implementing Micro-Hydro Power Systems

For those blessed with flowing waters meandering across your property, micro-hydropower may be an option.

Incorporating a micro-hydro system into your renewable energy plan allows you to tap into the power of moving water to generate clean, emission-free electricity. It's like completing the trifecta of renewable energy sources, alongside solar and wind power.

• Site Assessment Priorities

Before you get too excited about your micro-hydro dreams, it's important to do a thorough assessment of your water source. Just like with solar panels and wind turbines, the success of your micro-hydro project depends on a few key factors:

Flow rate: How much liquid volume constantly moves through your waterway over time? This quantifies your gross hydro-energy potential. The more water you have, the more potential energy you can harness.

Head (Vertical Drop): This is the total change in elevation as water flows through your property. The greater the drop, the more pressure you'll have to drive your turbine.

Elevation profile: Take a close look at where the water enters and exits your property, and how the water level changes throughout the year. This will help you figure out the best spots for your pipeline and water storage.

You might need to get your feet wet to gather all this information, but it's crucial for designing a system that works well with your specific site. By carefully measuring these hydrological patterns, you can start identifying the most practical micro-hydro equipment paths aligning with your specific needs.

- System Design Philosophy

With those foundational site studies done, it's time to start thinking about what kind of micro-hydro equipment you'll need.

This naturally begins by evaluating available turbine options based on your research findings. Different types of turbines work best for different water conditions. For example, if you have a low head (not much vertical drop) but a high flow rate, you might want to go with a compact helical turbine or an overshot waterwheel that can capture a wide swath of the water's energy. On the other hand, if you have a medium to high head but less water volume, a Pelton or Turgo impulse turbine might be a better fit.

The key involves thoughtfully selecting equipment honoring your source's unique hydraulics to maximize overall generation efficiency.

- Supporting Infrastructure

With your hydro-runner's mechanical heart selected, it's time to think about all the other components you'll need to make your micro-hydro system work. This includes things like:

Intake system: Vital for consistently capturing the water while filtering out debris that could damage downstream components. You might use a gravity-fed reservoir and pipeline, or a self-cleaning sediment chamber with screens to keep things flowing smoothly.

Penstock pipelines & electrical conduits: These are the pipes that will carry the water from your source to your turbine, and the cables that will transport the electricity from your turbine back to your home or battery bank.

Environmental harmonization: It's important to design your micro-hydro system in a way that minimizes any negative impacts on the environment. This means making sure you maintain a minimum water flow for fish and other aquatic life, and maybe even incorporating fish ladders or other structures to help them navigate around your system.

Energy storage & grid intertie: You'll need to think about how you'll store the energy your micro-hydro system generates, and how you'll connect it to your home's electrical system. Will you use batteries, or tie into the grid? Planning ahead will make it easier to expand your system in the future.

Admittedly, setting up a micro-hydro system on your property is no small feat. It takes time, money, and a lot of hard work.

But just imagine the satisfaction of knowing that every time you turn on a light or power up an appliance, you're using clean, renewable energy that you generated yourself! Not only will you be reducing your carbon footprint and reliance on fossil fuels, but you'll also be developing a deeper connection with the natural water cycles on your land. You'll start to see the landscape in a whole new way, reading the contours and flows like a map of opportunities for sustainable energy generation.

Energy Storage Solutions for Off-Grid Living

There's a catch-22 underlying this whole off-grid power puzzle—energy storage. If you can't store extra renewable energy when you have it and use it when you need it, all those renewable resources won't be very useful for your home's energy needs.

- Battery Basics

The basic technology enabling all this energetic squirreling away has existed for centuries in the humble rechargeable battery. But when it comes to storing renewable energy for your home, there's a lot more to consider than just the size and voltage.

Life cycle: How many times can the battery be charged and discharged before it starts to lose capacity?

Temperature tolerance: What temperatures can the battery handle without losing efficiency?

Discharge efficiency: How much of the stored energy can actually be used?

Battery chemistry: What materials are used in the battery (like lithium, antimony, or gel), and how does that affect performance?

It might seem overwhelming at first, but taking the time to understand these factors will help you choose a battery system that will last for the long haul.

- Load Calculation Foundations

To figure out how much battery storage you actually need, always first calculate your home's daily energy use. This means adding up the power consumption of all your appliances, lights, heating and cooling systems, and anything else that can plug in.

Once you have a total, it's a good idea to add a buffer of about 20-30% to account for battery capacity loss over time. The last thing you want is to run out of power when you need it most.

- Charge Management Wizardry

Unless you have an unlimited budget, you'll probably need to be strategic about how you charge and use your battery storage. This is where smart charge controllers and load prioritization come in.

For instance, programmable charge controllers continuously monitoring weather forecasts can strategically pre-charge backup reserves ahead of anticipated cloudy days or seasonal shortfalls. Or, if you have an electric vehicle with vehicle-to-grid capability, you can use your car battery to power your home during peak demand times.

Use technology to optimize your energy use and make the most of every precious electron.

- Backup Battery Bonanzas

Even with the best-laid plans, there may be times when your renewable energy system can't keep up with demand. That's where backup generators can be a lifesaver.

By integrating a fuel-powered generator into your off-grid setup, you can charge your batteries ahead of time and have a reliable backup source of power for emergencies. Some generators can even run on combustible waste like wood chips or biogas, turning trash into treasure!

The key is using smart charge control and prioritizing loads, automatically setting aside enough power for emergencies.

- Scalable System Integration

It's time to start piecing together a system that works for your unique needs and goals.; both today's realities and tomorrow's growth ambitions.

Maybe you're looking to power a remote cabin with a mix of solar panels and a small propane generator. Or perhaps you're part of a neighborhood micro-grid project that combines wind turbines, rooftop solar, and electric vehicle charging.

No matter your self-reliant endgame, thoughtfully engineered storage solutions demand holistic master planning around multi-modal energy routing equipment such as:

- Inverters to convert DC battery power to AC for your home appliances

  » Battery monitors to track performance and optimize charging

  » Smart circuits to prioritize essential loads during outages

  » And of course, a robust battery bank at the heart of it all

But there's more to it than just saving money and reducing emissions. By taking control of your energy production and storage, you'll learn to:

- Awaken before dawn to capitalize on those morning microbursts trickling into depleted reserves.

- Plan your daily chores and activities around the peak solar hours, when you have the most energy available.

- Share excess energy with your neighbors through small-scale community grids, fostering a sense of cooperation and resilience.

With your energy sources now under your belt, you can move on to the safety of your property, home, and family.

CHAPTER 6
FORTIFYING YOUR HOME: SECURITY AND PROTECTION

When you decide to go off-grid, you're not just saying goodbye to utility bills and city noise. You're also taking on the responsibility of keeping your family safe without relying on municipal emergency services or centralized infrastructure.

It might sound daunting, but with the right strategies and mindset, you can turn your off-grid homestead into a fortress of self-sufficiency and security. For now, we're going to explore a range of techniques that blend natural defenses, sustainable design principles, and smart technology to keep you and your loved ones protected and buy you peace of mind.

Natural and Sustainable Home Defense Techniques

Let's kick things off by tapping into one of nature's most underutilized security assets; your property's surrounding landscape!

Why settle for unsightly human-made barrier solutions when thoughtful ecological design can achieve equally effective defenses? You can use strategic plantings, natural obstacles, and biodiversity to create a living security system that's both effective and beautiful.

Principles of Biomorphic Security

The key is to think like nature does, using principles that have evolved over millions of years to protect plants and animals from threats. Some include:

Layered depth impedances: Just like dense rainforest underbrush makes it tricky to move around, creating dense shrubbery and hedgerows around your property's edges makes it hard for intruders to navigate and see what's inside.

Passive motion sensing: Many plants and fungi can detect movement and change their appearance or send out signals when they're disturbed. You can use these natural alarm systems to your advantage.

Resource masking: By cleverly hiding visible signs of occupancy (like cars or lighting) behind cleverly planted screens, you discourage unwanted surveillance and reduce the motivation for potential intruders.

Integrated firebreaks: Drought-resistant plants placed around your buildings can not only hide heat signatures but also create clear areas to defend against wildfires. Just ask those survivalists in Australia's bush communities!

Planting Protocols for Success

It's always good to think about creating three main layers of defense:

Exterior perimeter hedging: Start with a dense outer layer of thorny plants like hawthorns or barberries to deter human and animal intruders. These living walls can be both beautiful and formidable.

Interior obstacle banding: Add another layer of spiked plants like hollies or yuccas around your core living areas to slow down anyone who makes it past the perimeter. Choose native, aggressively growing species.

Central resource screening: In your inner sanctuary, plant edible groves, and fruit trees to hide resources and activities from prying eyes. You can even create bramble tunnels and trellised berries to blend your food production with hiding spots.

Each layer will reveal another hidden barrier and intruders will quickly realize how futile it is to try to breach your natural defenses.

Complementary Design Considerations

While cherishing botanical bodyguards, don't forget about the other features of your landscape that can enhance your security.

Earthworks and terrain contouring: Creating mounds, ditches, and land shapes can hide heat and make it tough for intruders to move. Pair these with ponds to obscure your ground presence even further and you have a winning recipe!

Maze-like domain planning: Creating small eco-communities with winding paths on your property also confuses and discourages trespassers.

Layering these natural defenses around your homestead provides an organic, interwoven protection system, where human structures complement rather than disrupt ecological connections. Given the choice between stumbling upon a menacing, fenced-off compound or a lush, vibrant garden paradise, which would you rather encounter on your evening walk?

Creating a Defensible Space Around Your Property

Any comprehensive security plan also needs to account for more severe external threats, like wildfires and floods.

So, what exactly constitutes an effective "defensible space" shielding your sanctuary from disastrous scenarios like wildfires, and floods?

A carefully designed buffer zone around your property that uses strategic landscaping and architectural elements to slow the spread of dangers and protect your main living areas against damage. Creating these protective buffer areas and using sustainable practices provides you with critical reaction time to activate emergency plans if the worst-case scenario becomes a reality.

Firewise landscaping

For off-gridders living in areas prone to wildfires, creating a fire-resistant defensible space should be a top priority. This means using smart landscaping and mitigation techniques, such as:

Fuel source reduction: Clear out combustible debris like dead foliage, dried grasses, and any other flammable materials within at least 30 feet of all structures. This starves creeping ground fires of fuel while buying you precious minutes. For bonus points, mulch the remaining wood scraps into eco-friendly ground cover.

Asset hardening: Speaking of exposed wood, upgrading all siding and deck materials to flame-resistant fiber cement or concrete eliminates easy ignition sources. Use metal roofing, double-paned windows, and enclosed undersides to prevent airborne embers from igniting your home.

Survivable landscapes: When it comes to new plantings, prioritize low combustion, drought-tolerant plants and arrange them in islands separated from buildings and woodlands. Use gravel pathways or stone features as natural firebreaks. Also, consider installing exterior sprinklers to keep high-risk areas hydrated during emergencies.

Topographical trenching: Some off-gridders dig sunken perimeter moats or rain gardens around inner buffer zones. These 4-foot trenches can trap spreading surface fires while also harvesting rainwater for firefighting reserves.

Compound Integrity

Environmental buffers serve purposes beyond inferno management. With some creative xeriscaping and habitat planning, these spaces can actually upgrade your overall compound's integrity through:

Erosion control: Carefully contouring perimeter terrain and planting stabilizers like nitrogen-fixing shrubs prevents soil erosion, mitigating slides or slumps during heavy rain. Cascading terraces with rainwater retention also stop flash floods.

Animal deterrence: By incorporating prickly barriers, like cacti or hawthorns, into your outer buffer matrix, you dissuade large mammal intrusions. Flowering pollinator-friendly patches provide safe forage too. It's all about balance!

High-traffic routing: Your buffer obviously shouldn't seal off access, so control entrances and exits by circumventing footpaths through greenways. Maybe that central roadway culminates in a secure gated checkpoint booth for visitor screenings?

Resource masking: Remember when cultivating interior food guilds, exterior buffers can discreetly screen activities and infrastructure from prying eyes too. Plant tiered hedgerows to obscure roof lines and blend outbuildings into the natural surroundings.

By thoughtfully designing these rugged defensive spaces around your off-grid homestead, you're not just enhancing your personal security, you're also cultivating environmental resilience and sustainability.

From the ancient Hanging Gardens of Babylon to traditional European hedgerows, gardenkeepers have always created havens of beauty and abundance amidst chaos. Now it's your turn to choreograph environmental defensive spaces that blend harmoniously with nature.

Advanced Surveillance and Alarm Systems

While Mother Nature is great at passively deterring threats, she's not exactly going to call for help if things go sideways. This is where high-tech electronics complement your eco-friendly paradise with 24/7 watchdog capabilities.

System Selection Factors

Between wired vs. wireless setups, AI-enhanced analytics, scalability, and compatibility with renewable power sources, it's easy to feel overwhelmed by all the choices.

That's why we'll break this subject down into key factors that should take priority:

Coverage requirements: First, calculate your property's total geographic footprint requiring surveillance. Do you need visual monitoring for large perimeters like farms or rangelands, or mostly for indoor spaces and outbuildings? Knowing your coverage needs will help narrow down your options.

Renewable power demands: With off-grid lifestyles in mind, always analyze each prospective system's total energy usage and accommodation for remote AC/DC switchovers. This will help eliminate potential wasteful electricity systems that could strain your renewable energy setup.

Analytics needs: Decide whether basic 24/7 motion-activated recording will suffice, or if you need more advanced features like intelligent behavior analysis that can flag specific threats like trespassing or vehicle breaches. Determine if simple or AI-driven situational awareness is right for you.

System architecture: Compare wired vs wireless network demands for reliability. Wired Ethernet systems offer consistent performance but require more setup effort, while wireless systems allow for easier scaling but may be vulnerable to radio interference.

By ranking your needs across these key areas, you'll narrow down available product options fitting your tailored off-grid needs.

Surveillance Centerpiece: Cameras

In most modern home security setups, cameras are the central situational awareness tool. And with today's wireless WiFi and mesh networking capabilities, achieving comprehensive coverage is surprisingly doable.

- For perimeters and open spaces, pair solar/battery-powered mast-mounted Pan/Tilt/Zoom (PTZ) turrets leveraging wide-angle thermal imaging alongside multi-sensor detection matrices, defeating concealment tactics. Intelligent motion tracking keeps targets centered too.

- Around structures, hardwired PoE doorbell cams interweave with wireless battery models. These thousand-yard stares overlook every nook issuing instant phone alerts on potential threats.

- Complementary alarm components like glass-break sensors or seismic vibration detectors can further enhance your primary camera systems, while AI-driven video analytics can customize threat notifications based on your specific criteria or behavioral anomalies. It's like having your own personal Big Brother!

Full-Range Monitoring

While cameras are essential, they're only part of the equation. By layering additional technological sentinels into your home defense matrix, you can strengthen your protection and eliminate coverage gaps:

Wireless sensor networks: Low-power, long-range (LoRaWAN) sensors powered by solar or battery can autonomously report conditions across your property. Strategically place moisture detectors, open/close indicators, or fence breach nodes to keep tabs on key areas.

Intercom communications: Two-way video intercom stations deployed at key property access checkpoints streamline secure visitor interactions, while passcodes and facial/vehicular recognition authorize entries for approved guests.

24/7 remote monitoring: Instead of constantly watching your surveillance feeds yourself, you can outsource monitoring to AI-powered offsite services that coordinate responses remotely. These intelligent analytics automate alerts while reducing false alarms.

Drone integrated viewing: Forget static cameras. The near future integrates autonomous drone patrols swarming airspaces on demand, beaming hyper-focused situational intelligence directly into your central security hub!

Perimeter Cameras Q & A

- Where should you put your security cameras?

    » Main entryways, like your front door, back door, and first-floor windows. This helps keep an eye on access points.

    » Perimeter corners and paths leading to main entrances.

    » Parking lots, garages, and areas with high-risk factors.

    » Avoid blind spots and ensure proper lighting for clear visuals.

    » Consider camera visibility to intruders and installation height to avoid tampering. You want to make sure they know they're being watched.

- What are all the basic building blocks of a CCTV system?

    » Commercial-grade cameras are suitable for indoor or outdoor use.

    » Structured cabling such as Cat5E or Cat6 for IP cameras.

    » Network Video Recorder (NVR) for storing and managing footage.

    » Hard drives for video storage and data retention.

    » Power-over-Ethernet (PoE) technology for efficient wiring.

- What types of security cameras can you choose from for your business?

    » Bullet cameras: visible deterrent, suitable for outdoor use.

    » Dome cameras: discreet, durable, ideal for indoor settings.

    » Turret cameras: versatile, good for low-light and repositioning.

    » PTZ (Pan Tilt Zoom) cameras: remote control, flexible viewing options.

    » Fisheye and multiple-sensor cameras: wide-angle coverage options.

- How to pick between DVR and NVR systems?

  » DVRs are suitable for existing coaxial wiring and analog cameras if you're old-school and proud!

  » NVRs offer higher resolution, compatibility with IP cameras, and remote viewing capabilities.

  » Consider your budget, existing infrastructure, and desired features for optimal choice.

- What storage options do you have for your security camera footage?

  » Onsite hard drives or Network Video Recorders (NVRs) for local storage.

  » Cloud-based storage for remote access and data backup.

  » Calculate storage needs based on resolution, bitrate, and retention period.

- What should you keep in mind when installing security cameras?

  » Proper cabling and wiring for reliable connectivity.

  » Protection against weather, vandalism, and tampering. A good rule of thumb is to install your cameras about 9 feet off the ground.

  » Strategic camera placement for optimal coverage and deterrence.

  » Professional installation for safety, compliance, and functionality. Sometimes it's best to leave it to the experts!

In these well-equipped spaces, your property becomes more than just a static fortress. It's a living, breathing ecosystem where natural and artificial defenses work together.

Emergency Plans and Communication Strategies

Even with Fort Knox-level security, not having emergency game plans and communication strategies in place leaves your entire self-sufficient dream high and dry.

Think of those nightmare situations catching even the most prepared off-gridders with their pants down. Maybe it's a raging wildfire sparked by some rogue lightning strike, or, a nearby industrial hazmat situation. Maybe this is the moment when those zombie apocalypse scenarios you've been nervously joking about actually start getting real! It's best to expect the unexpected.

Having pre-made emergency response plans tailored to your unique risks and resources means you won't be caught off guard when chaos strikes.

Building the Action Plan Blueprints

Where do you start when creating these emergency blueprints?

First, conduct a risk assessment to identify all the potential hazards or threats your homestead could face based on your location, geography, and nearby industrial sites. Do you live in an area prone to wildfires, floods, or severe weather like tornadoes? You'll want to prioritize plans for quick evacuation, emergency sheltering, and resource stockpiling. Regions at risk of earthquakes, hazmat incidents, or even terror threats need their own customized protocols too.

Once you've documented your locale's unique dangers and assigned probability scores to each one, you can start reverse-engineering the specific command structures, roles, and resources needed for an organized response.

The Core Emergency Blueprint

While precise execution details are situation-dependent, every comprehensive emergency action plan should incorporate some universal framework components:

Clearly defined roles & responsibilities: Who's in charge of getting the kids and pets to the safe room? Who's boarding up windows or setting up emergency power? Assign it all in advance, so there's no confusion when things get crazy.

Redundant communication protocols: Whether it's radio frequencies, satellite check-ins, coded text groups, or the classic hand-crank phone, your communication plan needs backups for when infrastructure fails. Everyone should know the primary, secondary, and backup channels for sending updates.

Rally points & evacuation routes: Before an emergency hits, map out designated interior safe rooms and exterior meeting areas. Stage 72-hour "get-home" bags with survival kits and make sure everyone knows the vehicle staging procedures. Run practice drills regularly, so no one's left behind.

Distributed supply caching: Don't keep all your essential stockpiles in one place. Instead, spread out your reserves of water, shelf-stable food, sanitation, and hygiene supplies, first aid, and prescription meds across multiple hardened caches on your property. That way, you're ready to bug in or bug out without scrambling.

Mutual aid partnerships: Some disasters will overwhelm even the most prepared individual homestead. That's why making agreements for mutual assistance with nearby homesteaders or community networks can be a lifesaver when things go sideways.

These action plan blueprints break down intimidating what-if scenarios into manageable, battle-tested procedures designed to keep your loved ones safe.

The Comms Resilience Backbone

The most meticulously engineered emergency plan is only as good as your team's ability to communicate and stay informed during infrastructure failures.

Start by designating a central command hub with redundant amateur radio systems, satellite data terminals, and landlines whenever possible. Tap into the power of self-healing mesh networks and ham radio integrations, boosting your team's wireless resilience with VHF/UHF repeater nodes and distributed gateways across your property. Robust GMRS handsets and laptop data hotspots can tie into these wireless internets, allowing coordination between on-site teams and remote helpers.

For worst-case scenarios, low-bandwidth emergency links like HF radio voice modems can keep digital comms going when terrestrial infrastructure totally fails. And if you're feeling old-school, dusting off analog line-of-sight signaling methods like semaphore relays or coded bonfires can be a good last-resort option too.

Fortifying your physical and technological defenses is important, but cultivating the inner resilience to face life's inevitable chaos is just as crucial. So, expect the unexpected, plan for the worst, and stay flexible!

CHAPTER 7
EMERGENCY PREPAREDNESS:
ESSENTIAL SKILLS AND KITS

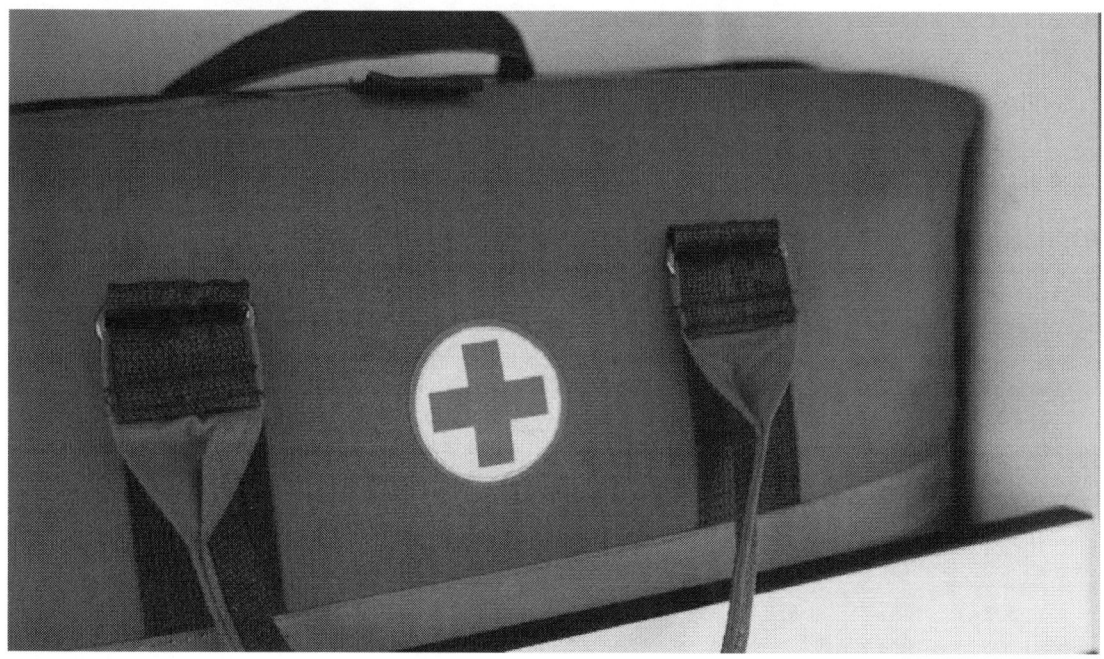

There's no point in sugar-coating this. Despite all our best-laid plans for renewable self-sufficiency, the cold hard truth remains that at some point, something may go utterly sideways. Maybe it's a cataclysmic natural disaster turning your scenic rural paradise into an apocalyptic hellscape overnight. One can never tell, and that's the point.

Comprehensive Guides to Assembling Survival Kits

Having a well-stocked collection of versatile survival kits tailored for various emergencies is an absolute must. Not exactly the greatest topic, right? However, nail these preparations and you'll suddenly possess a powerful self-reliance buffer for overcoming obstacles.

Basic Survival Kit

Here's a lovely little crash course in wound dressings and first aid kit essentials:

- What are sterile dressings used for?

  » Sterile dressings are like a protective shield for your wounds, keeping out germs and helping your body heal. Think of them as a clean, comfy environment for your wound to recover in peace.

- What types of wound dressings can you choose from?

  » Sterile wound dressings: These include sterile eye pads and other pads, perfect for keeping things clean.

  » Fabric and waterproof plasters: For small cuts or scrapes, these little guys have got you covered.

  » Adhesive dressings or plasters: From clear ones to blue catering plasters and even gel blister plasters.

- What's the difference between roller bandages and triangular bandages?

  » Roller bandages: These versatile bandages are like the Swiss Army knife of the first aid world. They can support joints, keep swelling in check, put pressure on wounds, and keep dressings in place. You've got conforming, open-weave, support, and self-adhesive bandages to choose from.

  » Triangular bandages: Made of cloth, these triangles can be folded into bandages or slings. And if you've got a sterile one, it can even serve as a dressing for bigger wounds and burns.

- What protective gear do you need for wound care?

  » Disposable gloves: When you're dressing wounds or dealing with bodily fluids and waste, these gloves are your new best friend. Go for the latex-free ones to avoid any unwanted allergic reactions.

  » Plastic face shields or pocket masks: These little lifesavers help prevent cross-infection when you're giving rescue breaths to a casualty.

- What other handy items should you have in your first aid kit?

- » Cleansing wipes: To clean the skin around a wound.
- » Gauze pads: For padding or a gentle wound cleaner.
- » Adhesive tape: To keep dressings and bandages in place.
- » Scissors: For cutting bandages, tape, or clothing to get to those hard-to-reach wounds.
- » Pins and clips: Used for bandage fastening.
- » Aluminum blanket: When you need to keep that body heat in check during an emergency.

Remember, when it comes to wounds, cleanliness is next to godliness.

Wilderness Survival Basics

For those of you settled in rural areas surrounded by the good untamed wilderness, a proper bushcraft bag should top your priority list. This sturdy backpack is built for thriving in the wild, with carefully organized gear covering key areas like:

Fire-starting: For fire through any conditions with waterproof matches, ferrocerium rods, fatwood bundles, and more. Be sure to pack multiple backup ignition methods.

Water procurement: Portable filters, Millbank bags, gravity filters, and maybe even a compact well pump to stay hydrated. Remember, running out of water is a death sentence.

Foraging & trapping: Bundle snares, fishing kits, and wild edible guidebooks for replenishing the initial 72-hour ration reserves.

Shelter and camping: Tarps, paracord, space blankets, and compact shovels that will keep you warm and dry to construct wilderness homes from scratch.

First aid kit: Bandages, medications, basic medical tools, and emergency response items. Include personal medications and emergency contact information.

Bushcraft basics: Stock lightweight stoves, multitools, cordage, and general-use pouches for improvising solutions to any unforeseen snags. Always include first aid kits, and the critical survival triad; fire-starting, edged tools, and signaling capabilities just in case rescue teams get involved.

Land navigation: Master compass reading, topographic charts, signal mirrors, and traditional land navigation skills because electronic GPS dependencies may not always work reliably during disasters.

Urban Escape Tactics

If you're pioneering off-grid living within large cities or surrounding suburbs, your kit needs some extra gear for hostile environment contingencies like civil unrest, chemical hazards, and subterranean navigation. Here are some key categories for staying mobile, discreet, and tactically flexible when bugging out in the concrete jungle.

Emergency sustainment: Discreet diaper bags containing consumables, cloaking camouflage, water filter straws, and accessories for quick relocation during rapidly deteriorating situations.

Cache resupplies: Pre-stashed resupply canisters along likely urban escape routes, containing ration blocks, ponchos, night vision optics, and physical maps for extended nomadic operations.

Counter-surveillance: Nondescript bags designed for anonymously blending into background city environments while transiting between safe areas. Think electromagnetic shielding, disguise makeup kits, and signal jammers for instance.

The name of the urban survival game? Never assume any escape scenario will unfold according to plan. Always build in multiple layers of backup when prepping for hostile situations.

Vehicle Sustenance Modules

Even casual road trips deserve emergency vehicle survival kits. Just in case that cross-country trip transforms into a multi-day detour.

At a minimum, stash these affordable trunk essentials.

- Jumper cables, emergency flares

- Lug wrenches and compact floor jacks

- Spare fuel tanks and flat repair supplies

- Blankets, ration blocks, water reserves

- Folding utility shovels and traction aids

- Reflective markers and signaling beacons

Don't forget to back up your mobile electronics with items like:

- Portable solar power banks

- Car inverters for radio/laptop use

- Backup GPS units and hardcopy maps

- Hand-crank/self-powered radios

- NOAA weather alert receivers

Nothing attracts fate's twisted sense of humor like neglecting basic emergency provisions. Better prepared than being the punchline.

Custom Kitted Configurations

I could spend all day listing specialized survival modules designed for every environment from aircraft to maritime to urban subterranean operations. But at the end of the day, putting together your own custom kits boils down to a few key principles.

- Conduct threat assessments in your area

- Group gear for likely emergencies

- Pre-pack for rapid deployment

- Regularly cross-train across all your equipment

Let the naysayers laugh at the "crazy prepper" stereotypes. True self-sufficiency means shedding that blissful ignorance and embracing nature's practical resourcefulness whenever you can.

First Aid and Natural Medicine Basics

What's the off-grid medic's mantra? Thoroughly mastering emergency first aid fundamentals mixed with natural medicine practices. You need to be able to handle any medical situation with the resources Mother Nature provides, at least until professional help arrives if it does at all.

First Aid Essentials

Establishing a solid foundation through basic conventional first aid training effectively buys precious time before fully-equipped evac teams arrive.

Understand important protocols like:

- Maintaining an airway, controlling bleeding, and preventing shock

- Properly cleaning and dressing injuries to avoid infection

- Immobilizing serious trauma like head/spinal/chest wounds

- Monitoring and treating environmental exposure risks

These skills might not be rocket science, but being able to perform them under pressure is what separates the cool-headed medics from the deer-in-headlights crowd.

Natural Remedy Integration

Mother Nature's medicine cabinet is fully stocked with plants that can tackle everything from cuts and scrapes to chronic pain. The trick is learning to identify, harvest, and use them properly. Start by cataloging the medicinal plants in your area, paying special attention to those with antibacterial, pain-relieving, or anti-inflammatory properties.

Get to know the different ways you can turn these botanical wonders into your own personal pharmacy, like:

- Poultices for stopping bleeding and promoting clotting

- Infusions and tinctures for managing fevers, coughs, and breathing troubles

- Salves for soothing burns, scrapes, and irritated skin

- Anti-inflammatory concoctions for reducing swelling around injuries

- Natural pain relievers for easing aches and pains, even chronic ones

Becoming a student of the healing cycles in your own backyard takes time and effort, but it's so worth it.

The Natural First Aid Kit

No respectable medic worth their weight aspires to roam the untamed wildernesses unprepared. That's why carefully provisioning a versatile mobile first aid kit stocked with both conventional emergency supplies and a variety of naturally sourced herbal and botanical remedies is important.

On the modern medical front, kits should include staple components like:

- Adhesive bandages, sterile gauze, and trauma dressings.

- Antiseptics, antibiotic ointment, wound irrigation tools.

- Blister care tools, splinting materials, elastic wraps.

- Scissors, tweezers, utility knives with extra blades.

- CPR shields, thermal blankets, activated charcoal.

But don't neglect to equip a separate "green farm-acy" section too loaded with:

- Organic disinfectants like colloidal silver gels.

- Peppermint, lavender, eucalyptus, or tea tree essential oils.

- Natural pain relievers like willow bark or turmeric powder.

- Echinacea, garlic, elder, or reishi mushroom antivirals.

- Wide spectrum herbal salves and tinctures like yarrow or propolis.

- Empty tins for making custom poultices in the field.

The Caregiver's Code

Being a self-reliant medic is about more than just having the right supplies and know-how. It's about forming a deep commitment to compassionate care and harm prevention in every aspect of your off-grid life.

Take a good look at the activities and routines in your neck of the woods and develop safety protocols like

- Making personal protective gear non-negotiable for any risky projects

- Using clear "challenge and confirmation" communication to avoid mix-ups

- Creating emergency code words to signal for help on the down-low

- Doing regular equipment checks and first aid drills to stay sharp

- Make sure your loved ones are trained properly on basic first aid and triage

When everyone comes together with mindful compassion, home becomes a safe haven, offering care and healing despite life's challenges.

Navigating Without GPS: Traditional Orientation Skills

One wrong turn could leave you hopelessly shivering in the wilderness begging for merciful death by toadstool.

Okay, maybe that's a bit dramatic, but the truth is, our reliance on electronic GPS has left many of us without the natural wayfinding instincts that our hunter-gatherer ancestors used to roam the land with ease. Reclaiming those skills, guided by the sun, stars, and natural landmarks, is the key to true self-reliance.

The Golden Orienter

We could deep dive into obscure archaic techniques involving sextants, mirrored astrolabes, or building sundial monuments from scratch. But for simplicity's sake, let's start with that free celestial sphere we all tend to take for granted; the sun.

By tracking the sun's path across the sky and its position relative to the cardinal directions, you can establish a remarkably accurate baseline for finding your way. Here are some key sun-based skills to master.

Rise and set tracking: The precise rising and setting points along horizon contours shift predictably both seasonally and latitudinally. So committing these patterns to memory as trekking references is a great start.

Solar noon positioning: Around midday, the sun reaches its maximum altitude directly above your true north-south line. Tracking the shadows cast by vertical objects at this time can help you find true north and south without a compass.

Watch recalibration: No off-grid timepiece keeps perfect time forever. But you can recalibrate your watch by tracking the progression of the sun's shadow on a makeshift sundial.

Starry Sojourning

Just like the sun, the stars and moon can guide you using ancient navigation methods. By combining these nighttime navigation techniques, you'll be able to find your way even when the sun goes down. Here's how stargazing can help you find your way.

The North Star: For

those in the northern hemisphere, the North Star (Polaris) sits directly above the North Pole, providing a constant reference point year-round.

Celestial sphere rotations: As the night sky rotates, the positions of the stars, planets, and moon can serve as a celestial clock and compass.

Star maps and stories: Many cultures have star maps that encode navigation cues, seasonal patterns, and even folk tales into the constellations. Rediscovering these ancient star stories can awaken your inner guide.

Constellation Chart

Constellation	Hemisphere	How to Find	Found
Ursa Major (Big Dipper)	Northern	Look for a group of bright stars resembling a ladle. Follow the two stars at the edge of the ladle's bowl to find Polaris, the North Star.	
Orion	Equatorial	Look for three bright stars in a row that form Orion's Belt. From there, you can locate other stars in the constellation.	
Cassiopeia	Northern	Look for a distinct "W" or "M" shape formed by five bright stars. It's opposite the Big Dipper and rotates around Polaris	

Constellation	Hemisphere	How to Find	Found
Scorpius	Southern	Look for a long-curved line of stars that resemble a scorpion's tail. Antares, a bright red star, marks its heart.	
Crux (Southern Cross)	Southern	Look for four bright stars forming a cross. It's one of the most recognizable constellations in the southern sky.	
Canis Major (Sirius)	Southern	Look for the brightest star in the night sky, Sirius. It's part of Canis Major and is visible from both hemispheres	

Natural Landmarks

For those times when you're too deep in the woods for sky navigation, developing a keen eye for local ecology and geography is just as important. The sights, sounds, and smells all around you can provide valuable clues to your location.

Terrain analysis: Every landscape feature, from the contours of the land to the flow of water and the shape of rock formations, can reveal your position if you know how to read them.

Plant migration mapping: Different plants and animals thrive in specific areas and follow predictable migration patterns. By observing the flora and fauna around you, you can often deduce your general location and heading.

Wind and weather pattern tracking: Changes in wind patterns, air pressure, and humidity can all provide clues to your location and help you predict upcoming weather conditions.

By practicing these skills regularly, you'll develop an intuitive sense of direction that will make you one with your environment.

The Pathfinder's Arsenal

Any skilled navigator knows that having the right tools can make all the difference. Some essentials include:

Analog Expedition Kits

- Compass (Lensatic/Tilt)

- Improvised sundial (using a hat, belt, or string)

- Sextant or clinometer

- Binoculars or telescope

- Celestial almanacs and star charts

- Paper maps and atlases

Ancestral Adaptations

- Dip bowls or water flotation compasses

- Wind maps etched on bark or stone markers (inukshuks)

- Lodestones and linear sundials

- Shell, stone, or seed chart (ronal)

- Walking sticks or staff

- Gnomon (shadow-casting) staff

Disaster-Specific Preparedness and Reaction Strategies

While you can't anticipate every possible scenario, gaming out the most likely disasters for your area can give you a critical advantage. No more getting caught off guard or reacting out of panic.

Natural Disaster Preparedness

For those of us living in areas prone to nature's fury, building specific contingencies into our seasonal routines is a must. This means having protocols for everything from emergency evacuations to long-term off-grid survival in the aftermath.

Wildfires: In wildfire country, maintaining defensible space around your home, having go-bags with respirators and fire shelters, and pre-staging emergency supplies are non-negotiable. Evacuation routes, fire blankets, and sprinkler system backups should be covered. Leave nothing to chance.

Earthquakes: Those blessed with tectonically active environments should prioritize surveying structural vulnerabilities, anchoring entry, and exit contingencies, and always keeping water and food stockpiles rotated. Don't forget to plan out emergency communication methods and radio ranges.

Tornadoes and hurricanes: When tempest winds and storm pressures pose constant threats, having a designated household storm shelter or reinforced safe room is crucial. Make sure to restock your first aid kits and emergency supplies before each storm season, and consider building a backyard storm cellar for extra protection.

Flooding: Those inhabiting floodplains should strongly consider raising utilities, reinforcing stem wall foundations, and even building site relocation if seasonally exposed. Emergency dewatering pumps in low areas will help prevent washing away.

The common threads? Have pre-planned evacuation routes and off-site meeting points, and practice your emergency drills regularly.

Human-Catalyzed Emergencies

Not all disasters come from Mother Nature. Sometimes, the biggest threats come from human activities or malicious intent. Here's how to prepare for some common man-made emergencies.

Industrial accidents: Living within potential blast radiuses from refineries, rail yards, or manufacturing zones demands contingencies like emergency decontamination showers, eye and respiratory protection gear, and even radiation detectors. Hazmat isolation procedures should also not be overlooked.

Societal breakdowns: Those anticipating potential civic unrest fueled by economic collapse or resource wars, rehearsing self-quarantine and self-defense protocols while hardening perimeter security through concealment or evasion, should definitely make the checklist.

Environmental emergencies: In areas at risk of industrial pollution or toxic spills, having environmental monitoring equipment like contaminated runoff alarms near your water sources can give you a critical head start on mitigating the damage.

Tactical incursions: While none of us ever want to actually experience them, common sense still demands at least entertaining defense strategization for hostile scenarios. Identify escape routes, pre-position survival caches, and maintain your self-defense training. It's not paranoia, it's preparedness.

Most importantly, have multiple layers of communication backups to coordinate with your loved ones, emergency responders, and community support networks.

Holistic Reaction Flowcharts

A good rule of thumb is to create simple, step-by-step flowcharts for each type of emergency. These checklists will take the guesswork out of keeping your family safe when every second counts.

Some reaction checklist templates worth developing in advance include:

Initial emergency assessment matrix

- Quickly identify hazard type, threat level, and potential impact radius.

- Use tools like emergency alert apps, hazard sensors, and weather radar to stay informed.

Primary response procedures chart

- Establish your top priorities, like getting everyone to a safe place, activating your shelter protocols, and staging your emergency gear.

- Document the locations of your supply caches, family meeting points, and vehicle staging areas.

Communication flow chart

- Plan out multiple ways to stay in touch with your team, both on-site and off-site.

- Use pre-established codes, encryptions, and check-in protocols to keep everyone informed as the situation develops.

Evacuation procedure maps

- Have at least two escape routes planned out, along with provisions for clean water and food along the way.

- Practice your procedures for securing your home, coordinating transportation, and checking in at your destination.

Extended off-grid contingency plans

- Clearly outline your capabilities for food, water, energy, and sanitation if you need to bug in for an extended period.

- Establish mutual aid agreements and resource-sharing plans with your neighbors and community to weather the storm together.

Recovery checklists

- Once the dust settles, have a framework for assessing the damage, rebuilding what's broken, and getting back to a new normal.

- Prioritize safety, environmental cleanup, and updating your disaster prep plans based on lessons learned.

At its core, disaster preparedness is all about taking personal responsibility for your own safety and resilience. You can't always rely on others to swoop in and save the day, and you can't afford to sleepwalk through life ignoring the fragility of the systems we depend on.

CHAPTER 8
DIY OFF-GRID SHELTER: FROM PLANNING TO CONSTRUCTION

You've got your off-grid lifestyle all planned out, but you still need a sustainable space you can call home. While the information in this chapter may seem somewhat ambitious, building your own self-sustainable home can be greatly rewarding. And, if you're not planning on building anything from scratch, don't skip over this chapter as it will provide you with tips and tools to help upgrade your current home too!

Designing Your Off-Grid Home:
Principles and Considerations

Let's start off by laying some philosophical groundwork first.

If you're building an off-grid home, you're not just looking for a weekend vacay spot. You want to create a space that's in tune with the natural environment, a place where your daily life and the surrounding ecosystem can coexist in harmony.

To achieve this, you need to embrace a few core off-grid design principles:

- Blend your home harmoniously with the landscape.

- Use locally sourced, eco-friendly materials where possible.

- Create closed-loop, sustainable systems for water, waste, and energy, to minimize external dependencies and maximize sustainability.

The word is biomimicry! This means that you take cues from nature's patterns and incorporate them into your own designs. It's all about working with the natural rhythms of your surroundings.

Foundational Maxims

Keep the following key principles in mind for home harmony with the environment.

Prioritize Passive Design

- Use a smart, climate-responsive design that harnesses natural light, airflow, and thermal mass by orienting your home to take advantage of the seasonal sun and wind patterns. Use and sculpt the land itself, creating natural air and water flow.

Emphasize Local or Regional Materials

- Use local, natural building materials that best blend in with the colors and textures of the environment.

Decentralize Utilities Integration

- Create closed-loop systems for water, waste, and energy. Renewable energy like solar, wind, and hydropower as well as incorporating greywater recycling and composting toilets will maximize sustainability.

Cultivate Multifunctionality

- Create multi-purpose spaces that seamlessly blend living and working for better flow and functionality. Hone your permaculture skills to integrate your home with food production and natural ecosystems like implementing composting and water harvesting systems.

Plan for Modular/Scalable Adaptability

- Design your home to be modular and adaptable, using movable walls, adjustable shelving, and other flexible features to maximize your space.

Conduct Environmental Assessments

- Identify all unique characteristics, resources, and challenges to adapt and design accordingly by conducting thorough environmental assessments.

Contemporary Inspiration Repositories

There are vast communities of off-grid pioneers out there, paving the way for sustainable living. Some great educational resources and communities include:

Permaculture design courses and collectives: Great for learning how to create self-sustaining ecosystems that integrate human habitats with natural abundance.

Passive house and regenerative design standards: Get the "know-how" on green building techniques like passive solar, geothermal, and earth-sheltered design.

Earthship biotecture and natural building schools: Most certainly worth looking into for off-grid building methods that use recycled and natural materials in innovative ways.

Indigenous construction collectives: Great for insights into traditional building techniques that have been used for centuries to create sustainable, locally adapted homes.

University extension programs: Expand your knowledge of local ecology, sustainable agriculture, and rural living best practices.

Always blend your human living patterns with nature's rhythms as much as possible to not clash with them.

Natural Building Materials and Techniques

Another epic topic: natural building materials!

From bricks browned and baked in the sun to breathing bamboo, there's a whole world of sustainable options out there!

Natural Building Materials

A few natural construction heavy-hitters include:

- Earthbag/Bagcalcrete

You've probably noticed those half-buried "hobbit homes" constructed using a technique called earthbag or Bagcalcrete building.

This method is one of today's most popular off-grid crazes for good reason. The process involves stacking and compacting courses of sturdy polypropylene sandbags filled with an insulative earthen plaster mix. These adaptable structures are compacted and covered with mounds of earth, sprayable concrete, or vegetation for weather protection.

Earthbags are cement-free and mainly sourced from local soil excavations or quarries, making use of local materials like clays, sands, and fibers.

- Cob Construction

Then we have cob; a traditional construction method using earthen materials.

Made from highly saturated, sandy mud, it generally consists of subsoil clays, fibrous organic matter like straw, coarse sand aggregates, and simple stabilizers. By continuously massaging and kneading the soil mixtures, the cob transforms into sturdy monolithic walls with excellent thermal mass properties.

- Straw Bales

For the more adventurous folks, consider straw bales. This involves stacking and compressing conventional straw bales into dense, load-bearing monolithic walls, exhibiting incredible insulative properties.

Those bale "logs" essentially become enormous compacted air pockets, just like double-paned windows. Except, these are entire foot-thick, passive, climate control walls that are fire-resistant too. Efficiency at its finest. Mastering straw bale construction involves learning about essential details such as strong baling twines, breathable stucco for moisture control, and ensuring weight is properly distributed across ceilings and foundations.

Material Sourcing Strategy

Before getting swept up in romantic notions of sculpting cob cottages or straw bale havens, you need to address some material sourcing strategies.

Environmental surveys: Conduct surveys of your site to identify any potential hazards and contaminants. Tip: steer clear of uranium hotspots!

Sustainable supply mapping: Identify nearby sources of natural building materials, and consider the logistics of transporting them to your site. Make sure the soils, rocks, and plants in your area are durable and suitable for construction. Keep an eye out for renewable sources of plant fibers, like straw and reeds, that can be harvested without depleting ecosystems.

Community relationships: Expand your knowledge and experience by connecting with artisans, builders, and suppliers who specialize in natural building materials.

Material Overviews

Some of the most promising natural building materials to consider include:

- ## *Compressed Earth Blocks*

Compressed earth blocks (CEBs) are the modern, mechanized version of traditional mudbrick. Made with a mechanical press, these blocks provide extra strength and density, ideal for hot, arid climates where moisture isn't a problem. However, it does require stabilizers like Portland cement or lime binders for enhanced water resistance.

- ## *Light Straw-Clay*

This is a mixture of loose straw and clay, used to create insulative, breathable wall panels. This moldable light, straw-clay easily sculpts wall-integrated shelving and other components while regulating temperature. It makes it ideal for humid climates where moisture control is a factor.

- ## *Bamboo & Timber Framing*

Bamboo's incredible strength, rivaling steel, combined with timber's flexibility generates sturdy framing systems, aligning beautifully with ground motion. Fill in the frames with earthen masonry or other natural materials for insulation, and you have yourself an environmental winner.

- ## *Ferrocement & Papercrete*

These are more modern cutting-edge techniques that employ thin, strong shells using wire mesh and cement plaster, or a mixture of recycled paper and cement. The result is significant structural integrity and impact resistance. Similarly, papercrete replaces Portland cement with recycled paper fiber, turning waste into functional, insulated walls.

The beauty of natural building materials is that they emerge directly from the resources available around you, creating a home that's truly in tune with its surroundings. Just remember, natural construction goes beyond building a structure; always balance technical details with mindfulness when choosing building materials.

Thermal Mass Heating and Cooling Solutions

Let's take a closer look at the power of bricks, stone, concrete, and even packed earth to soak up the sun's heat and slowly release it, keeping your home toasty in the winter and cool in the summer. That's right! You don't have to be a hostage of utility companies anymore. It's just you, your thermal mass, and a solid connection to nature.

The Mass Concept

The idea behind thermal mass is simple—certain heavy, dense materials are really good at absorbing, storing, and slowly releasing heat.

This includes materials like bricks, stone, concrete, adobe, and even tightly packed earthbags or rammed earth. In the summer, they absorb excess heat from inside your home during the day and release it outside at night, keeping you cool and comfy. Pair this thermal storage power with some smart passive solar design, and your home becomes a self-regulating climate control machine.

Strategic Choreography

To unleash your thermal mass's full potential, you'll need to account for a few things.

Solar orientation: This is rather obvious! Strategically position your thermal mass surfaces, like concrete floors or brick walls, for optimal daily sun exposure. South-facing windows are your friend!

Material qualities: Not all dense materials are created equal when it comes to thermal mass. Evaluate metrics, like density, specific heat capacity, and thermal conductivity, to pinpoint top performers. For instance, concrete and masonry generally outperform lightweight wood framing's heat retention.

Glazing balance: You need to strike the right balance between the amount of thermal mass and the amount of window glazing. Too much glass and not enough mass; your home will heat up and cool down too quickly. Too much mass and not enough glass; you'll end up with a slow, heat-soaked cave.

Airflow management: To keep the stored heat flowing where you want it, you need to be smart about ventilation. To control how heat moves in and out of your home, use operable windows, vents, and adjustable shading, like overhangs or awnings

Mass Solutions

You'll also need to tailor your system to your specific home design, climate, and energy setup. Some solid solutions include.

Trombe walls: These heavyweights use a south-facing glass wall in front of a high-mass concrete or masonry wall to soak up the sun's heat during the day and slowly release it into your living space at night. Add some smart vents, and you've got a passive heating and cooling system.

Roof pond systems: Water bladders are sandwiched between your roof's exterior surface and an interior thermal mass ceiling below. In summer, these bladders cool the ceilings and radiate heat skyward at night. In winter, the drained bladders allow solar heat to pass through, providing passive warmth.

Masonry heaters: These high-mass wood stoves are like the Swiss Army knife of heating. They burn a small, hot fire that slowly heats up a maze of brick or stone chambers, radiating gentle warmth into your living space for hours after the fire's gone out.

Earth tubes: For earth-sheltered or underground homes, innovative ventilation ducting can help leverage Earth's subterranean thermal mass for passive climate control. You get cool air intake during the summer, and warmer fresh air during the winter.

Hydronic radiant: By circulating hot water through a network of tubes embedded in a thermal mass floor, like concrete or brick, you can turn your entire floor into a heating panel. Reverse the flow in the summer, and you've got a built-in cooling system too!

Instead of fighting against nature with brute force, you're working with the natural rhythms of the sun and Earth to keep your home comfortable year-round.

It's more than just about saving money. You are turning your home into a living, breathing extension of the natural world, rather than a sealed-off box.

For a classic example, take a peek at the brilliance of a Trombe Wall system. You've got south-facing glazing strategically separated from an interior high-mass wall, like concrete or stone, with just enough air cavity between them to let the magic happen. During winter days with low-angled sunrays, the light penetrates right through the glazing, saturating the masonry wall with thermal energy throughout the afternoon's optimal heating period. Then once outdoor temperatures drop, all the stored heat inside the wall slowly releases inwards, distributing heat into your living spaces overnight.

Take the traditional Russian masonry stove heating network. A relatively small, controlled wood fire travels through a complex indoor masonry flue system before exiting at a stove outlet to disperse the heat. The gentle smoldering fires gradually saturate all the baked-in thermal mass with retained warmth first. That thermal battery then trickle-releases heat for between 12-24 hours, even after extinguishing the firebox.

So whether you're building a sun-soaked Trombe wall, a cozy masonry heater, or an earth-regulated hobbit hole, you're leaning into true off-grid climate freedom.

Water and Sewage Systems

Unless you're aiming to become a smelly off-grid recluse, any legitimate homestead requires implementing comprehensive water supply and waste treatment systems.

In this section we're taking the plunge, talking about rainwater harvesting, greywater recycling, blackwater management, and even pulling water out of the air. You need to get smart about managing the water cycle from start to finish.

No plumbing enigmas or civil engineering voodoo is needed, just a willingness to harmonize with hydro-flows.

Securing Your Water Supply

Priority numero uno: securing a reliable water source to feed your homestead with adequate volumes. This means taking a close look at your local rainfall patterns, groundwater resources, and humidity.

Collecting rainwater from rooftops into storage tanks is often the easiest way to get water. Just make sure you size your storage tanks properly for your local wet and dry seasons and don't forget to filter out any debris and treat the water before you drink it. If you're lucky enough to have a well or a spring on your property, directly tap into that groundwater goodness. Do your homework on the local hydrology and water table first, and don't start pumping without a sustainable plan in place. For folks in ultra-thirsty climates where water options are limited to atmospheric humidity or occasional fog, using thermal condensers with reverse osmosis filters can create sustainable water from humidity. Be warned though; those atmospheric chillers require some serious energy inputs in return.

So what is your best bet? Diversify your water sources whenever possible. Relying on a single supply is like putting all your eggs in one leaky basket.

Wastewater 101

Any water wizards worth their salt understand that recovering fresh water only solves half of the hydrological puzzle. Dealing with wastewater is just as important as finding fresh water! The difference between greywater and blackwater?

> **Greywater:** This water refers to the lightly tainted streams from showers, sinks, and washing machines, free of solids and contamination. With some simple filtration, settling to remove sediment, and possibly using natural methods like constructed wetlands, this water makes an ideal underground irrigation supply.

Don't go chugging that recycled dishwater, because greywater can contain harsh chemicals and other contaminants. Always keep your greywater pipes separate from your drinking water, and stick to eco-friendly soaps and cleaners where possible.

Blackwater: This water contains heavy bioloads straight from toilets, kitchen scraps, and stables, dripping with bacteria and concentrated nutrients. It's too potent for direct recycling without rigorous multi-stage treatment.

Composting toilets are a great way to deal with the solid waste, turning poop into rich garden compost. However, not for your edible patches. You'll need a multi-stage treatment system that uses a mix of anaerobic digestion, aeration, and filtration to break down the sludge and clean up the water. Just remember; blackwater demands obsessive microbiology-level treatment gymnastics before redistributing it back towards potable quality. Depending on local laws and regulations, you may need to jump through some hoops to get your blackwater system approved.

Legalities and Compliance

No amount of off-grid plumbing manifests legally, or sustainably without navigating a maze of municipal code restrictions, public health ordinances, and environmental regulations.

Most government jurisdictions lay down strict water quality monitoring and treatment standards you'll need to comply with. Acute pathogen risks like Giardia get prioritized over chronic chemical pollutants. Rainwater catchments often face unique stagnation and set-back restrictions to navigate.

Similarly, segregating wastewater streams, pretreatment protocols, and final disposal all face scrutiny across the board. From handling stormwater runoff and securing sewer conveyance permits to addressing limits on groundwater injection, managing biosolids can be quite complex.

Thankfully, more pragmatic "Water Reuse Categories" and "Risk-Based Frameworks" have emerged, alleviating some regulations. Additionally, color-coding makes it easy to tell apart lower-quality greywater from higher-quality effluents by looking at potential risks. So thoroughly plan your water management, covering risk prevention, readiness for emergencies, and reuse strategies in your permit applications.

Installation Wizardry

Now that we've covered the basic, bureaucratic fine print, let's get your hands dirty and make your off-grid hydro-sanitation masterpiece a reality with the following essential steps:

1. Rainwater collection: Calculate rooftop surface areas and local precipitation rates to properly size your wet/dry cistern capacities. Make sure to include filters and overflow outlets for initial treatment.

2. Well installation: Survey groundwater levels and gradients to determine well depths, casing requirements, and sustainable pump specs. Don't blow your water budget with an overpowered pump!

3. Stream/greywater segregation: Install p-traps and sewer vents to separate concentrated waste from reusable greywater streams. Avoid cross-connections at all costs.

4. Primary treatment: Basic grease traps in kitchen drains and traditional septic tanks can begin eliminating solid waste and fats, oils, and grease (FOG) buildup immediately.

5. Greywater distribution: From there, valve manifolds, drip irrigation lines, and raised feeder pipes distribute greywater to underground drip areas in the landscape, where it nourishes plants effectively.

6. Blackwater bioremediation: To treat strong wastewater effectively, you'll need to set up specific systems like anaerobic baffled bioreactors or aerobic activated sludge sequencing batch reactors. These systems break down, purify, and refine the wastewater for potential reuse.

This process closes the nutrient loop, maximizing your self-sufficiency. One cycle's effluent becomes another flow's influence over and over, making every drop count. From diverting rainshower catchment to irrigating kitchen gardens, which evaporates to atmospheric humidity and clouds once more. Those greywater streams filter through botanical biotreaters sprouting veggies to eat. Blackwater sludge solids compost down into fertile soil. It all comes full circle!

CHAPTER 9
COMMUNITY BUILDING AND NETWORKING FOR RESILIENCE

The idea of going full-on hermetic and living entirely off your own regenerative capabilities holds great appeal, I'm not going to lie!

Not relying on anyone, living off your own land. It all sounds pretty tempting, right? But even for the most skilled and prepared off-gridders, trying to be completely self-reliant as a one-person show would be like playing a video game with no extra lives. You're choosing to survive without any chances to start over. The far savvier move? Becoming a passionate advocate for building self-reliant local communities and fostering resilience through collective empowerment. By networking with nearby homesteaders and nurturing resource-sharing groups, your dreams of self-sufficiency get a boost through the power of cooperation. It's all about strength in numbers!

What resilient off-grid homestead couldn't benefit from trusted allies pooling talents, exchanging surpluses, and coordinating mutual safeguards?

Forming and Strengthening Local Networks

You've read the manifestos, watched the survival flicks, and daydreamed about that solitary life foraging fungi in some remote Rocky Mountain hideaway far from mainstream society.

Unless you're actively dead-set on being a complete hermit, plugging into local self-reliance networks and creating resilient communities around you might just be the most powerful move you can make for long-term sustainability. Think about it for a second. Even if you manage to achieve that dream of living off wild-harvested food, what happens when you get seriously hurt miles away from help? Or when seasonal changes disrupt the delicate balance of resources you depend on? One unpredictable hiccup and poof! There goes your entire self-contained off-grid vision!

You're going to need to factor in trusted allies, combining their skills and resources in cooperative communities that promote sharing. These communities can span across small neighborhoods or even small towns. With community support and sharing, you can weather just about any storm thrown your way. Always remember that communities generally thrive, especially when compared to isolated forms of self-preservation where you only survive.

The Local Networking Game Plan

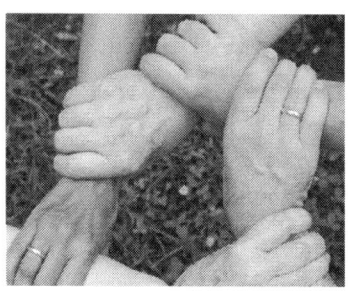

Where do you even begin to build these crucial connections?

Luckily, igniting local self-reliance communities from scratch, or reviving slumbering ones, proves far more straightforward than you'd imagine. Some basic principles you can follow include:

Broadcast your passion: Start sharing your love for self-reliance, cooperative resilience, and decentralized sustainability across all the social channels you're part of. From community bulletins to neighborhood apps, social media channels, or just evangelizing friends and family. Amplify that shining passion of yours whenever you can!

Identify like-minded spirits: Keep an attentive eye peeled for any kindred spirits whose self-expressed values, hobbies, or lifestyles might indicate compatible mindsets. Passionate permaculturists? Hard-core DIY enthusiasts? Closet off-gridders quietly prepping below radar? These all make prime potential allies worth reaching out to.

Host meet-and-greets: With a handful of eager visionaries aboard, start hosting casual community meet-and-greets centered around self-reliance topics. You could host workshops, seed swaps, potlucks, or sustainability lectures to name a few. Always create a welcoming, no-pressure atmosphere that brings newcomers together through shared interests.

Gauging Shared Interests

As relationships develop, thoughtfully assess areas where networking participants have particularly strong shared interests. Permaculture gardening? Renewable energy projects? Identifying shared focuses naturally illuminates springboards for launching collaborative initiatives everyone can support.

• Proposing Collective Endeavors

Once key shared interests become clear, start energizing the crew by pitching resilience-building community projects, tapping into everyone's potential. You can create crowdsource funding and labor for cooperative infrastructure uplifts like

• installing neighborhood food forests or CSAs (Community Supported Agriculture programs).

  » developing shared seed libraries and nurseries.

  » solar array co-ops and microgrid power installations.

  » educational workshops teaching survival skills.

  » establishing formalized neighborhood watch programs.

  » coordinating disaster prep networks and drills.

  » organizing barter marketplaces for goods and services.

The real beauty is that collaborative projects multiply participation, diversify contributions, and keep resources circulating in closed regenerative loops that ultimately facilitate a resilient group of close-knit people.

- Formalizing Structure

Speaking of formalization, once initiatives gain traction, thoughtful governance structures preventing conflicts become valuable.

Collectively develop mission statements, operating principles, and boundaries, and draft fair resource-sharing agreements that incentivize active commitment. A bit of democracy helps harmonize cooperative efforts without compromising independent sovereignty too much.

- Cross-Pollinating Connections

Never become the proverbial big fish trapped in a small self-reliant pond. Just as established networks start flourishing locally, actively cross-pollinate those connections so that you can cultivate relationships with allied resilience communities on broader regional scales too.

Resource coordinators, skill instructors, specialized suppliers, you name it. Expanding complementary channels between groups widens access to invaluable capabilities that exponentially amplify everyone's resilience. Who knows, maybe someday entire self-sufficient locally-governed bioregions could emerge?

Realize that your self-reliance revolution is really just the beginning, not some final isolated destination. Even our self-sufficient hunter-gatherer ancestors deeply understood the symbiotic group dynamics of sustainability long before the rise of agriculture and settled living. Trying to function as a stubborn self-contained loner back then simply meant a short lifespan!

Self-reliance may have begun with boot-strapped individual pursuits. But real sustainable resilience only thrives cooperatively within thriving empowered communities.

Collaborative Projects for Shared Resources

Simply surrounding yourself with like-minded sustainability enthusiasts isn't quite enough to maximize a collaborative spirit's full potential.

The savvy move would be to actively lead ambitious shared resource projects that directly leverage your group's pooled expertise, volunteer labor, and interconnected asset streams for mutual empowerment. It's all about strategically harnessing the "strength in numbers" advantage to create regenerative abundance cooperatives that go beyond what any individual could achieve alone.

Some solid collaborative endeavors along these lines might include neighborhood-scale installations like:

• *Tool & Equipment Libraries*

Getting all the specialized tools and power equipment necessary for comprehensive off-grid self-reliance over time requires serious upfront capital that most of us simply don't have lying around. What if there existed a community-owned cache of shared gear that covers everything, from industrial-grade preservation equipment to excavators, that any participating household could borrow as needed? Sounds good, right? Funding one premium chainsaw suddenly becomes more feasible!

• *Seed Banks and Nurseries*

A similar logic applies to cultivating agricultural seed independence. Instead of each homestead maintaining separate seed vaults with multiple climate-specific varieties, jointly coordinating neighborhood-scale community seed archives, germination facilities, and plant nursery co-ops enables far more comprehensive biodiversity. Those regionally-adapted seed varieties get lovingly preserved and shared throughout these networks.

• *Resource Recovery Networks*

Collaboration in sharing and recycling resources is also important for efficient resource use. Neighborhood recyclers, material reuse cooperatives, food rescue initiatives redistributing grocers' unblemished surpluses, and even innovative community composting hubs all demonstrate value-added renewability. Perhaps your collective could pool funds to construct a neighborhood anaerobic digester installation, capturing food and yard waste streams, generating renewable biogas, and producing nutrient-rich soil amendments for community gardens. Symbiotic energy and nutrient cycling possibilities are truly endless.

• *Renewable Energy Co-Ops*

By pooling interest from participating households, perhaps your community can crowdfund a solar micro-grid setup. Or you could commission developers to launch small-scale distributed wind or micro hydro projects, strengthening energy democracy. On an even more hyper-local level, innovative solar thermal installations delivering hot water, radiant heating, or crop dehydration services become more attainable through collective resource pooling. The same goes for establishing shared reinforced storm shelters, emergency preparedness reserves, or designated crisis communication infrastructures.

• *Managing Shared Stewardships*

Structured governance cooperatives with elected boards and fair resource-sharing principles help harmonize collective efforts behind unified value propositions. Digital ledgers that transparently track participation credits and duties are invaluable for equitably allocating surpluses while discouraging disengagement. On-site operational frameworks outlining facility access procedures, borrowing protocols, and basic maintenance responsibilities are make-or-break considerations as well. Some basic democracy prevents the inevitable personality conflicts.

Genuine self-reliant resilience doesn't necessarily mandate individual resource ownership. Or to put it another way; does owning your personal chainsaw really unlock more value than having access to a shared tool cache managed by the community, which can fulfill almost any homesteading requirement?

Embrace the power of collaboration and lead efforts in your community to share resources and promote resilience together.

Developing a Barter Economy:
Exchanging Goods and Services

Ready to get down and dirty with some good old-fashioned wheelin' and dealin'? It's time to resurrect one of humanity's most ancient traditions—barter economies!

What is barter anyway? Bartering simply involves directly trading goods and services between two parties without involving any form of conventional currency as an intermediary transaction medium. It's the "you scratch my back, I'll scratch yours" model of mutual aid. Our ancestors embraced this very concept before the powers-that-be started dictating what little pieces of paper were supposed to be worth. Barter economies bring back rationality by trading resources directly, preserving community balance with sensible exchange rates.

Cooperative Benefits

But enough history for now. What are the benefits of bartering?

For starters, ditching centralized currency dependencies outright eliminates financial exploitation from leeching institutions and their inflationary devaluation schemes entirely. When you cut out the middleman and deal directly with your neighbors, you're taking back control of your economic destiny.

By keeping resources flowing in a circular pattern within your local ecosystem, you're not just reducing waste and strengthening community bonds, you're creating a regenerative feedback loop that just keeps on giving. Surplus homegrown produce swaps for handyperson construction labor, which in turn trades for preserving the same nutritious foods long-term, and so forth.

It's a practice that reconnects you with the true value of our labor and resources. When you trade your homegrown goods or skills directly with someone else, you can't help but feel a profound sense of pride, gratitude, and interdependence.

Local Barter Frameworks

With all those cooperative benefits singing their song, how do you actually go about setting up a thriving swap economy?

The most common methodology involves establishing a local community exchange network and coordinating barter listings and trade ratios amongst participating members. These days, straightforward apps and websites can streamline entire virtual back-end experiences, from cataloging inventories to facilitating trades seamlessly. More ambitious collectives may even establish physical barter marketplaces, hosting swap meets where folks gather and trade from preserves and plantlings through to metalcraft and repairs.

Managing Mediums & Values

Inevitably though, larger barter economies might need to get a little creative to keep things running smoothly.

Maybe you introduce some kind of community currency or time-banking system to help facilitate more complex trades or come up with a clever way to store and preserve surplus goods for lean times. The key is to stay flexible and communicative, and always prioritize the needs of the collective over any one individual's hoarding tendencies.

Resource Balancing

Let's keep this barter party rolling with the delicate art of resource balancing!

The thing about a thriving barter economy is that it's always in a state of dynamic equilibrium, but it also takes a bit of finesse and constant adjustment to keep everything humming along smoothly.

The last thing you want is for one particular resource or skill set to become so overrepresented that it throws the whole system off. Imagine a scenario where everyone and their mother is suddenly hoarding the same seeds while basic homestead maintenance tasks start piling up. Before you know it, you've got a bunch of seed-rich but infrastructure-poor preppers.

An elegant return to environmental harmony occurs when one person's prosperity nurtures another's flourishing, creating a cooperative and regenerative resonance that multiplies outward like fractals.

That's where the magic of community-driven value assessments and proxy adjustments comes in. By regularly gathering as a group to take stock of your collective resources and needs, you can nab potential imbalances and keep the barter train running smoothly. Is there a sudden glut of artisanal candles flooding the market? Time to temporarily boost the value of more essential goods and services to encourage a little redistribution of wealth, for instance. But resource balancing isn't just about playing with supply and demand, it's about creating a resilient, regenerative foundation. That means investing in the kind of infrastructure that allows you to store, preserve, and manage your collective bounty with the changing of the seasons.

Creating a system that's built on abundance and shared prosperity, enables something magical to unfold. People start to realize that true wealth isn't about hoarding resources or clawing your way to the top of some imaginary ladder, it's about cultivating a sense of interdependence and mutual growth.

When individual passions and skill sets start to harmonize and flow together, well, that's when the real regenerative magic starts to happen. A barter economy that's firing on all cylinders is a network of reciprocal exchanges and feedback loops that creates a rising tide of abundance for everyone involved. And isn't that what this whole off-grid rebellion is really about when you get right down to it? Rejecting the scarcity mindset and artificial limitations of the status quo, and embracing a more organic, holistic way of relating to each other and the world around us.

If we learn to see our individual gifts and resources not as something to be hoarded and guarded, but as a sacred offering to be shared freely for the greater good of the collective, that's when we start to glimpse the true meaning of regenerative wealth.

CHAPTER 10
SUSTAINABLE LIVING BEYOND BASICS: ADVANCED PROJECTS

The cutting-edge sustainable living scene means non-stop innovation. If you're not constantly leveling up your game, you're gonna get left in the dust faster than a biodegradable plastic.

The common thread that ties all these eco-innovations together like a perfectly woven hammock is about embracing a holistic lifestyle; every last scrap of resources is optimized to the max, and waste is just a dirty five-letter word.

Automating Off-Grid Systems for Efficiency

I've got an intriguing proposition that might just take your self-sufficiency passions to exciting new fronts: why not consider strategically integrating some cutting-edge automation technologies into your hard-earned self-reliant infrastructure? What if you could have the best of both worlds?

Before you start grumbling about potential contradictions between environmental philosophies and encroaching AI overlord agendas, we're not talking about outsourcing your entire off-grid homestead's operations to a soulless robot battalion. Instead, I'm suggesting surgical strikes of smart tech in all the right places. Think precision agriculture sensors for your permaculture garden, or aquaponic systems that basically turn fish poop into veggie gold.

Imagine instead of spending hours tweaking your irrigation system or balancing your aquarium chemistry, you could just kick back and let the algorithms do the heavy lifting.

Automated Efficiency Examples

Are you still skeptical about how technology could ever harmonize with your beloved off-grid ethos? Well, let's take a quick peek at some areas ripe for integrating efficiency-enhancing smart gadgetry.

Precision agriculture: What if strategically deploying a constellation of remote IoT sensor nodes scattered across your permaculture gardens could stream real-time data on soil moisture, temperature, and nutrient levels directly to customized software? It could be done! This software also automates your irrigation schedules and fertilizer applications down to the square foot. Eliminating annoying guesswork while slashing resource waste.

Aquaponics systems: Imagine a self-contained ecosystem where fish waste feeds hydroponically grown veggies, which in turn filters the water for the fish in a perfectly balanced nutrient cycle. Now add in some smart sensors and automated dosing systems, and you've got precise management of fish waste and hydroponic plant growth without needing constant manual adjustments.

Smart power management: Over in the residential renewable energy area, you can let the algorithms do the heavy lifting. Smart power management automatically load-balances between different power sources, prioritizing critical loads during outages, and even predicting your energy needs based on weather forecasts and usage patterns.

Utility automation: Let's not forget the simple quality-of-life benefits smart home automation unlocks throughout utility and security domains. Envision automated lighting schedules, thermostatic climate control, smart locks and cameras with facial recognition convenience, and automated gray/blackwater recycling systems, making your daily chores a whole lot easier.

The Smart Setup Process

Now that you're starting to glimpse the exciting cross-pollination potentials between old-school, self-reliant philosophies, and new-school technological optimizations, you're probably wondering what's even involved with integrating them anyway.

Here's the basic gameplan:

It involves conducting an extensive home energy audit and calculating anticipated load profiles that will account for that new smart tech's electricity too. Based on those assessments, you'll have to properly size residential solar systems, battery reserves, and any supplemental generator backups needed.

Next, evaluate foundational automation platforms like comprehensive smart hubs supporting protocols like Z-Wave, ZigBee, and voice control abilities, all internet-connected within your home automation ecosystem.

Then comes the fun part, integrating various specialized smart device peripherals. Water quality monitors driving hydroponic dosing. Environmental sensors automating climate control and growing lighting. Surveillance cameras, smart locks, automated shades and blinds, you name it.

Finally, reinforce network resilience through cybersecurity audits and rigorous system backups/redundancies. After all, self-reliance means never completely outsourcing residential independence. Not even technologically!

When you do automation right, it's not about giving up control—it's about taking control back from the daily grind and freeing up your time and energy for the things that really matter.

Keep an open mind regarding automated assistance possibilities. Imagine a future where every off-grid homestead is a perfectly optimized ecosystem, with smart tech and natural processes working together in perfect harmony to create a truly sustainable way of life.

That's the kind of future that gets me excited!

Developing Sustainable Aquaponics Systems

Let's see how to turn a mouth-watering fantasy into eco-reality through the incredible horticultural ingenuity known as aquaponics.

It's all about seamlessly combining aspects of aquaculture (raising fish) with hydroponics (soilless plant cultivation) into one harmoniously closed food loop system, operating with minimal environmental impacts. An aquaponics system enables nutrient exchanges between three core components symbiotically fueling each other. Waste byproducts that fish expel get naturally filtered by bacteria colonies, whose resulting nutrient-rich water then flows right into hydroponic growing beds to feed a thriving produce ecosystem before recycling condensates back through aquaculture tanks again.

My description barely does justice to capturing how aquaponics actualizes those elusive aspirations towards waste-free, self-sustaining food independence. So, let's dive a little deeper into some design considerations and components that make these systems such modern agricultural wizardries.

The Design Philosophy

Before even browsing fish species or seedling selections, any prospective aquaponics artisan needs to thoughtfully map out overall system parameters like:

Production volume: Calculating anticipated yields tailored for individual or family consumption versus commercial scale operations. This will enable you to gauge how you'll need to scale things.

> **Energy budgeting:** Don't underestimate power appetites from pumps, aerators, grow lights, and environmental controls. Renewable backups and energy optimizations form an essential part.

Climate control: Temperature, humidity, and air circulation all heavily impact growth rates and overall system efficiency. Plan for managing microclimates across all your components.

Spatial footprints: While compact aquaponics units do exist, comprehensive systems tend to command fairly sizable floor spaces, accommodating all their interlinked components.

Living Component Selections

The real fun begins when selecting which specific aquaculture and hydroponic species works symbiotically best for your situation. Factors worth weighing include:

Fish types: Hardy species like tilapia or bluegill thrive in diverse water quality ranges while providing ample nutrients for plants to thrive on. But more delicate specialties could add commercial value.

Hydroponic crops: Leafy greens, herbs, and vegetables lacking extensive root systems and tolerating relatively high nutrient levels make great hydroponic matches for fish fertilization.

Supplementation needs: Don't underestimate potential mineral, oxygen, or pH supplementation requirements for thriving environments. Planning buffering systems early will prevent costly nutrient deficiencies.

The Operational Choreography

With all system components calibrated, maintaining an aquaponics installation involves meticulously monitoring and actively optimizing that balanced nutrient cycle between fish, microbe colonies, and plant ecosystems.

- For instance, sludge buildup in fish tanks must get routinely cleared so ammonia levels don't spike toxically. However, over-clarifying those biosolids could deprive hydroponic vegetable beds of their primary fertilizer source at the same time.

- Similarly, regulating water chemistry across interconnected components remains absolutely crucial for stable environments.

- Nitrite/nitrate concentrations must convert efficiently for microbes and plants. Slight pH fluctuations quickly stress biological communities disrupting entire aquaponic equilibriums.

- Preventative reservoir aeration, environmental controls, and judicious supplementation are of utmost importance.

The smart solution? Advanced sensors, automatic dosing, and smart climate control systems efficiently manage nutrient distribution in aquaponics, making it more accessible than ever.

But for those thinking about expanding aquaponics beyond home setups, there's potential to inspire whole neighborhoods to become self-sufficient in agriculture. Imagine collaboratives launching community-scale aquaponic projects, selling fresh produce locally, and reinvesting profits to grow into self-sustaining food sources for the area. Even more exciting? Wastewater from neighboring homes could simultaneously provide nutrient-dense fish food inputs while receiving reclaimed freshwater, completing sustainable cycles.

Crafting and Smithing:
Utilizing Resources Fully

Perhaps in our haste to streamline self-reliance through technological prowess, we've accidentally left behind certain timeless, quintessentially human creative pursuits. Ancient hands-on talents, connecting us with material flow through humble artistry, not circuitry!

Allow me to reintroduce the often-overlooked arena known as traditional crafting and smithing. Those deliberately slow disciplines resurrect the "lost souls" of material resources through mindful ingenuity, shaping nature's raw abundance into purposeful, household implements, cherished artworks, and more. These arts of crafting and smithing beautifully embody the zero-waste ethos through continually reimagining and repurposing locally available elements into utilities before our very eyes. Every end is another rebirth.

The Material Canvas

But exactly what sorts of raw matter might contemporary crafters channel their ingenious resourcefulness toward? The better question is what local or household wastes can't be transformed into custom goods instead.

Plant fibers: Fronds, leaves, woods, and grasses can weave into wonderfully diverse textiles and decorative baskets. Bamboo can be crafted into practically any household furniture piece imaginable.

Clays and earthworks: Locally excavated soil clays sculpted through mindful cobb, ceramic, or raku firing present endless possibilities suited for anything from dishware to outdoor sculptures.

Metal alloys: Got scrap metals lying around? Aluminum cans, copper wiring, and steel can all be liquified into jewelry, tools, or bells. Just plan proper safety protocols!

Repurposed Salvage

The only true crafting limit is when we fail to see the essence of each element before reshaping it. Every material deserves appreciation through the lens of creation.

• Timeless Techniques

Mastering material ingenuities all begins through disciplined technique training. Don't worry too much about robotics replacing these time-honored instructors yet. What artisan apprentice wouldn't salivate learning skills like:

Textile weaving: Mastering knot-work patterns or plaiting fibers into fabrics, rugs, or ribbons through looms and braiders will enhance your skills in textile arts.

Woodworking: Basic sewing, carving, steam-bending, wood joinery, and lathe spindling will unlock boundless home furnishing possibilities.

Metallurgy: Perfecting your crucibles, molds, alloy compositions, and casting/smithing forms will polish your blacksmith skills.

Ceramic: Wheel-throwing, extruding, under/overglaze decorating, and expert kiln firing will have you creating more than cups and collectibles.

The meditative repetition linking body movements with every stitch or pour combines momentary inspiration with lasting results.

The Regrowth Movement

Vibrant off-grid talent hubs are popping up all around the world. Within these inspired craft-centric communities, rich cultural traditions fuse ancestral ingenuities with progressive eco-mindsets.

Cooperative members barter skills and resources, kids apprentice under master clothiers and ceramists, and market days celebrate zero-waste artworks. You will see renewable abundance transform into elegant furnishings, wearable textiles, or functional home goods. These thriving crafting ecosystems are like a cultural rebirth, reconnecting us with ancient materials and unlocking endless ingenuity. Inner creativity and outer abundance are finally in harmony without any conflict.

So regardless of whether you envision mastering production pottery or metalsmithing, always remember that the physical acts completing material cycles simultaneously perfect your spiritual fulfillment too.

Innovative Recycling and Waste Management Practices

If we collectively keep treating basic household "waste" as an inevitable burden headed straight for the local landfill, then we're still fundamentally missing something crucial, aren't we?

Transcending those outdated disposal mentalities, counting every used plastic container or food scrap as burdensome trash is a "no-go." Nothing featuring resource potential should ever be carelessly discarded. Every perceived "waste" stream merely represents another misidentified opportunity waiting for your innovative repurposing genius. The big picture goal here involves pragmatically finessing those aspirational "Zero Waste" and "Circular Economy" ideals into practical closed-loop systems, cutting out wasteful linear consumption completely. An existence where absolutely nothing gets taken for granted anymore, and every perceived "waste" output simply feeds another complementary production input.

DIY Zero Waste Projects

Let's jump straight into highlighting some super inspiring zero-waste strategies you can start implementing throughout your self-reliant homestead:

Household composting: Elevate basic backyard compost heaps into vermi-composting, flow-through reactors, processing food scraps, shredded papers, leaves, and manures into nutrient-dense soil for organic food production.

Greywater recycling: Let nutrient cycles complete themselves naturally. Domestic greywater from sinks, showers, and washing machines can be redirected into botanical growth beds, flourishing with edible landscaping.

Waste plastic processing: Why toss synthetic polymers into some distant pit when they're such versatile feedstocks instead? Shredding, melting, and remanufacturing waste plastics into new household goods completes the loop perfectly.

Waste mycelium construction: Mycelium-based biocomposites can create insulation panels, acoustic tiles, and even structural furniture components from self-replicating fungi that digest woody and cellulosic waste streams into building materials.

Biochar production systems: Pyrolysis reactors can transform biomass "wastes" into incredibly nutrient-dense biochar fertilizers, boomeranging right back into your permaculture gardens. Now we're talking closed loops.

These represent just the tip of the proverbial iceberg. Once you embrace that zero waste paradigm, suddenly your entire home morphs into a perpetual closed-loop bioreactor. Everything gets upcycled, recycled, or regeneratively repurposed before exiting the premises.

Domestic Nutrients Cycling

You can start integrating anaerobic digesters to break down blackwater sewage into mineral-rich soil amendments, nurturing food forests. Or, maybe install solar stills harvesting atmospheric humidity into pure condensation streams.

Essentially, when you retrain your mindset recognizing "waste" simply as untapped resource potentials, suddenly sustainable ecosystem dynamics replace linear consumption patterns entirely. Your household's hydration, energy, and nutrition needs get self-supplied through regenerative resource flows.

• The Zero Waste Gameplan

Exactly what facilitates transitioning any household into an ultra-efficient closed-loop resource recovery station? Here's a simple game plan to kickstart your zero-waste revolution:

Audit every conventional "waste" stream currently exiting your homestead.

Keep track of what you have and the resources you might be wasting without realizing it.

With the inefficiencies quantified, scour innovative case studies, implementing accommodating recycling/upcycling technologies for recovering value from waste flows. Just make yourself ready for some inspirational brainstorming.

Develop custom strategies to capture the resources available in your off-grid environment more effectively. Think greywater recyclers, plastic reclaimers, bokashi fermenters, Myco-construction clusters, and nutrient concentrators.

Integrating well-coordinated utility processes helps transform each recovered "waste" output into inputs that fuel other regenerative systems. This integration rcduces resource demands as household operations become more interconnected.

Living in true self-sufficient harmony with nature necessitates transcending these outdated "disposal" paradigms altogether.

Everything cycles back into complementary recovery pathways, ensuring nothing goes to waste. Your role is to creatively choreograph these regenerative flows throughout living spaces, optimizing sustainability.

Fully embrace these ethics because in consciously liberating every previously discarded "waste" product into its highest value, you'll be harnessing your fullest regenerative potential.

CHAPTER 11
LEGAL AND FINANCIAL PLANNING FOR OFF-GRID LIVING

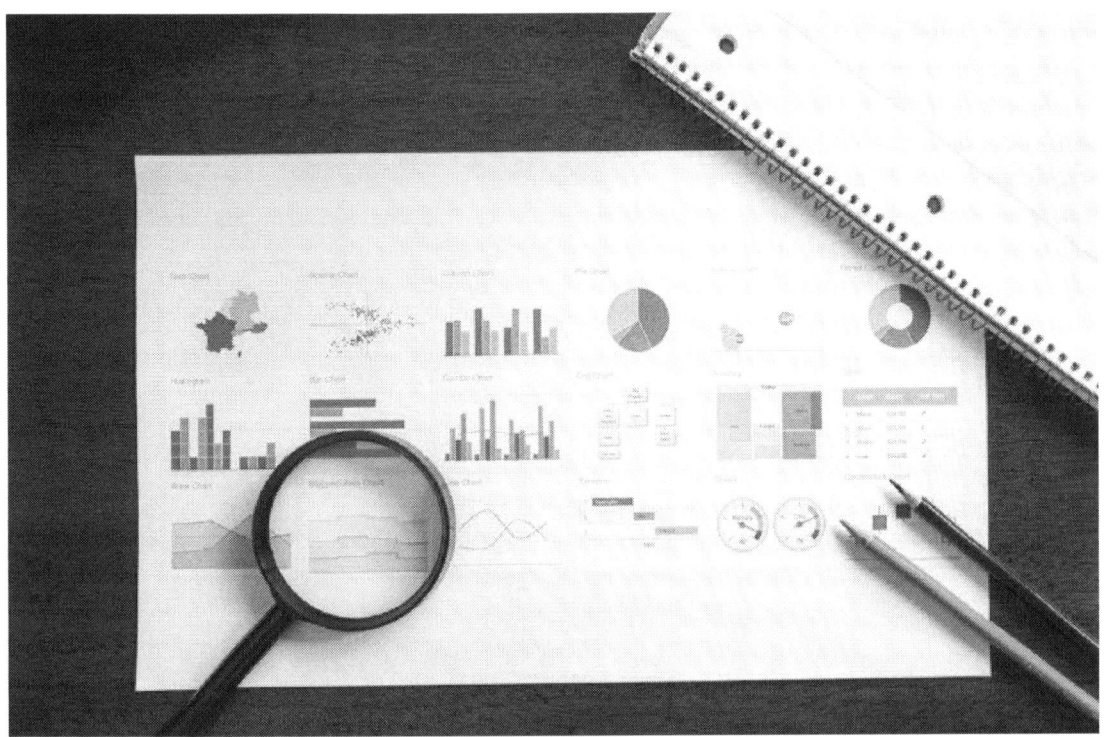

We've covered all sorts of self-reliance disciplines so far. From sustainable construction and renewable energy generation to closed-loop food production and hardcore zero-waste. But unless you're already a multi-millionaire squatting in international waters, chances are your path towards that covetable independent living dream still faces one particular gatekeeper—legal bureaucracy!

Even the most ecologically-attuned, high-tech self-sufficient homestead unfortunately can't just manifest itself out of nowhere without cautiously dotting all those pesky compliance i's and crossing regulatory t's first. I'm referring to the unavoidable exhaustive zoning assessments, environmental impact analyses, building permit acquisitions, and mountains of nitpicky code conformances still standing between your current reality and that liberated off-grid vision. But don't sweat it. We'll be unpacking those red tape rigmaroles.

And hey, since we're already preparing for the administrative anarchy of bureaucracy, why not get ahead of the financial planning game too? After all, funding such an ambitious, regenerative refuge requires meticulous budgeting and crafty resource allocation.

Navigating Zoning Laws and Building Permits

If you've stuck with me this far, it's probably safe to assume you're pretty serious about making this whole off-grid living thing an achieved reality rather than just an idealistic pipe dream.

We're about to step into arguably the most soul-crushing, profanity-inducing, tear-out-your-hair, infuriating aspect of manifesting that self-reliant dream: navigating zoning regulations and permitting processes. Yes, we have to run the full bureaucratic gauntlet against an endless battalion of nitpicky zoning officials, building inspectors, and heavy-handed enforcement officers, each armed to the teeth with dense legalese designed to obliterate your sovereign dreams before they even leave the drafting table. Alright, perhaps I'm exaggerating a tiny bit, but, there's still some truth in what I'm saying.

If you're truly committed to doing this whole off-grid thing legally and above board, there's simply no way around eventually squaring off against the zoning and permitting establishment's finest. It's a harrowing inevitability you must stoically prepare for.

Research and Reconnaissance

How does one properly steel themselves for such regulatory warfare? The first crucial maneuver involves conducting extensive scouting missions across the battlefields you'll be campaigning within, namely, your local municipal zoning codes and ordinances.

Many aspiring off-gridders dream of creating an ecological haven in rural areas. This often means carefully navigating county-level zoning rules that dictate land use for areas like agriculture, residential estates, or conservation districts. Important details to scrutinize here include:

- Lot size and setback requirements

- Permitted construction practices

- Limits on activities like agricultural operations or home-based businesses

Urban homesteaders face their own set of challenges in navigating municipal codes, which often focus on strict regulations for low-density residential or industrial zones. Anticipate scrutinizing regulations concerning:

- Property usage

- Environmental impacts

- Utility provisions

- Neighbor complaint statutes

No matter which zoning theater your game unfolds within, missing even one stealthy zoning rule can disrupt your plans and permitting efforts. So do your homework diligently!

Securing Approvals

Next comes the real harrowing test; actually petitioning for formal construction approvals. Let's just say it can be a grueling process.

Core permit submissions typically demand marshaling comprehensive dossiers packed with construction documents (site plans, engineering details, etc.), environmental impact reports, surveyed plats, and even contractor licensing verification before they begrudgingly open a case file. Miss any paperwork requirements, and the entire application gets stamped "Denied" faster than you can blink. Then, after countless reviews and intense questioning by city officials, you might just snag that golden ticket of approval for your project.

Don't forget, permit fees often necessitate their own financial planning as well. Many zoning authorities sadistically calculate staggering surcharges for residential builds based on parameters like total square footage, property acreage, or estimated private utility loads. Budget accordingly!

The Advocacy Badger

Navigating zoning codes can be painfully slow, especially when they don't keep up with modern off-grid living ideas. Persistence is often the key to success in such situations.

This may entail attending public zoning board hearings, badgering commissioners with custom proposal justifications, or even recruiting community support campaigns behind sympathetic civic leaders. Hey, nobody ever promised that spearheading self-sufficiency revolutions would feel breezy.

Keep trusting your instincts, as authorities also have limited enforcement powers. Many proactive zoning officials actually welcome sustainable living additions, enriching stale municipal landscapes nowadays, if diplomatically pitched. Some might even support updating outdated rules ahead of time.

Keep moving forward patiently, nurturing flexibility alongside principles, and reframing obstacles as opportunities for advocacy wherever viable.

Financial Planning and Budgeting for Off-Grid Projects

How exactly do off-gridders strategically finance pulling off such ambitious homesteading endeavors without depleting every last penny into money pits along the way?

We've got to confront some harsh economic realities here and put some cold, hard facts on the table. Even the most brilliantly intentioned dreams of solar-powered aquaponic abundances backed by rammed earth masonry require intensive financial forecasting and meticulous resource allocations. Otherwise, that half-built straw bale homestead quickly devolves into just another abandoned, eyesore construction site, cluttering rural roadsides sooner than you think! Let's demystify the personal financial circus accompanying serious off-grid undertakings.

Off-Grid Costing Fundamentals

Before even broaching potential funding sources or juggling budgetary line items, any dedicated off-gridder must first cultivate an honest grasp on just how weighty those comprehensive implementation price tags could balloon to.

For larger-scale off-grid homesteads blending natural construction techniques with supplemental solar arrays and water autonomy systems, initial costing exercises should cover categories like:

Land acquisition expenses: Beyond just purchasing raw acreage itself, factor zoning and legal fees, surveying contracts, environmental assessments, and even procurement logistics.

Residential construction budget: From permitted civil site work through to vertical construction materials and labor, meticulously scope even finishing details like cabinetry or tilework.

Renewable utility provisions: Scrutinize renewables such as solar and battery storage installations, well-drilling with auxiliary water treatment systems, septic and greywater recycling, and backup generator gear.

Agricultural establishment: Never neglect comprehensive provisioning costs like animal pens and shelters, crop cultivation zones, post-harvest processing, and preservation equipment.

Project Financing Options

Up next, securing project financing commitments and underwriting those expensive off-grid fantasies into reality.

Personal asset liquidation: To fund off-grid projects, consider selling assets like investments, property, or business shares. Ensure you clear debts and secure enough funds after expenses.

Traditional lending solutions: More conventional financing routes also remain available through established institutional lending streams. Prospective mortgage lenders, home equity loans, peer-to-peer funds, and lines of credit should all get comparatively scrutinized by interest rates and terms.

Public and private partnerships: Those pursuing more ambitious community-scale off-grid undertakings may engage public agencies, anchoring grants or government-subsidized loan guarantees, underwriting green initiative costs. Similarly, private investors seeking tax credits or alternative asset portfolio diversification represent viable funding options as well.

Crowdsourcing campaigns: For others, orchestrating aggressive equity crowdfunding campaigns through online platforms, and soliciting broad-based community investment audiences could viably bridge funding gaps. Just be meticulous in fulfilling promised deliverables and prospectus obligations.

Naturally, most comprehensive projects embark on combining diverse financing and allocating appropriate proportions through staged capital structuring to optimize cost efficiencies.

Cost Management Disciplines

Even the most thoroughly financed off-grid endeavors still need careful budget management to maintain financial integrity during setup and operation.

Efficient procurement practices: Always strategically source the highest-quality materials and services at the lowest possible costs through competitive bidding processes, incentivizing contractors while negotiating volume purchasing agreements. Scrutinize invoices hawkishly too.

Logistics and overhead optimizations: Judiciously managing labor productivity, supply chain efficiencies, and administrative leanness to streamline overhead costs goes a long way. Even small recurring costs amplify alarmingly fast.

Iterative budgeting reviews: It's absolutely crucial to conduct routine operational budget forecasting, and analyzing actual spending patterns versus initial projections. Dynamically recalibrate discrepancies, immediately mitigating cash flow discrepancies before crises snowball.

These are just some key basics to help you get the ball rolling in the right direction. At the end of the day, there's no sugar-coating reality; flawless financial management is the key to keeping ambitious self-reliant goals on track and avoiding financial mistakes. So take those money matters seriously from visioning through perpetual operation. Stay principled, stay disciplined, and stay lucrative.

Insurance Considerations and Risk Management

Your collective resilience repertoire is rapidly reaching sustainable levels!

Despite absorbing all these invaluable skills for liberating yourself from industrial society's overreliance on centralized systems, we're left with the final critical consideration: how exactly do thoroughly self-reliant homesteaders protect their priceless ecological sanctuaries from inevitable risk eventualities?

Things like natural disasters, electrical hazards, accidents, or legal liabilities could send all your hard-earned self-sufficiency crumbling before your very eyes. The not-so-glamorous yet absolutely mandatory world of off-grid insurance and residential risk management strategies is up next.

The reality is that once you embark on the independent homesteader's path, you essentially take on comprehensive custodianship over protecting your paradise from any conceivable eventualities.

Residential Property Coverage

Let's start by outlining some core residential property insurance essentials, providing dependable coverage for your off-grid homestead's physical assets and liabilities. Categories worth evaluating include:

Dwelling protection: This ensures the reconstruction value of your residence itself, but many providers offer extensions covering detached structures like barns, sheds, renewable energy systems, and fences.

Contents coverage: Don't forget to insure everything inside your home; furnishings, appliances, personal possessions, and gear. You'll likely need supplemental personal property coverage.

Liability protection: No policy feels complete without mitigating legal liabilities, should visitors suffer injuries or trespassers initiate lawsuits related to your property. Liability coverage shields your personal assets.

For more adventurous setups, certain supplemental riders provide added risk mitigation.

Agricultural operations: If operating farms, gardens, or livestock, you'll need extra coverage beyond basic residential policies to comprehensively protect all your productive assets.

Environmental insurance: Progressive eco-friendly construction or land stewardship measures may qualify you for incentivized "environmental" policies from some providers. Be sure to ask.

Off-grid endorsements: As sustainable living grows, customizable industry endorsements are emerging to holistically cover households generating their own renewable power or food. What a winner!

Your best bet is to consult with insurance agents or brokers specializing in alternative energy, sustainable living, or off-grid properties. These professionals provide valuable guidance and options tailored to the unique needs and challenges of off-grid living, ensuring adequate coverage for properties, equipment, and potential liabilities.

Additionally, join forums or communities of off-grid enthusiasts and seek recommendations from experienced individuals to help find the best insurance providers or advisors for off-grid living.

Risk Analysis and Assessment

Doesn't sovereignty inherently reject dependencies on corporate authorities promising theoretical safety nets in exchange for perpetual costs?

Yes and no. Divorcing those dependencies doesn't negate the harsh reality that disruptive events could still impact even the most robust self-reliant oasis. Principled self-sufficiency transcends total isolation but absolutely demands comprehensively self-managing unavoidable risk exposures versus recklessly ignoring them.

That's why leveraging private insurance solutions represents a remarkably pragmatic way to facilitate residual risk transfers covering those inevitable "what-ifs." But it begins with rigorously quantifying potential threats to your assets first:

Threat cataloging: Systematically document every conceivable hazard scenario that could impact operations; fires, storms, equipment failures, legal disputes, casualty events, etc.

Probability modeling: Leverage data on historical events, environmental patterns, and actuarial analytics to gauge each threat's relative likelihood annually. Prioritize high probabilities accordingly.

Impact forecasting: For each identified threat, project worst-case physical, financial, and legal impacts on infrastructure, populations, and economic continuity. Quantify potential losses.

Mitigation planning: With prioritized shortlists, develop strategies to reduce threat probabilities or limit consequence severities wherever achievable through prevention and planning.

With outstanding risk exposures quantified, you can then facilitate transfers through strategic policy acquisition balancing premiums, deductibles, and coverage limits appropriately. Facing reality remains far cheaper than tempting chaos.

Policy Portfolio Management

Even the most meticulously planned insurance regimen demands ongoing vigilance to maintain coverage relevance. Policies require routine maintenance just like organic gardens!

Asset reviews: Regularly check that all your home assets, equipment, and tasks are well-planned to avoid any gaps or issues over time.

Regulatory compliance: Make sure your coverage meets the latest safety codes, laws, and credentialing needs set by authorities, and update policies promptly.

Endorsement inspections: Continually review additional endorsements, replacement cost terms, or liability limit adjustments as your operations evolve or grow.

Carrier rotations: Never settle for complacency! Strategically renegotiating placements between underwriters incentivizes competition while optimizing coverage quality.

Independence's path winds at times, but safeguarding your self-actualized paradise's longevity is immensely worthwhile!

Insurance Intersections

Developing a robust risk management strategy for your off-grid homestead inevitably intersects with some of the legal and financial disciplines we've explored:

Zoning and construction: Confirming structural renovations or civil infrastructures comply with applicable building codes prevents coverage gaps during claims. Permits matter!

Agricultural operations: Innovative integrated homesteads combining crop or livestock farming with value-added processing require specialized agribusiness insurance packages.

Renewable utilities: As residential solar arrays, wind turbines, and off-grid power systems grow mainstream, unique equipment coverages are required too.

Business pursuits: For homesteaders monetizing surplus yields through farm stands or cottage industries, comprehensive commercial policies protect those entrepreneurial endeavors.

Estate planning: Coordinating your risk management strategies with comprehensive estate plans protects family legacies, spanning inheritances and trust arrangements.

The path toward authentic self-reliant independence ultimately demands diligent risk mastery across legal, financial, and operational domains alike. So, analyze, mitigate, insure, and maintain!

CONCLUSION

More profound than the tangible technical skills gained, I hope this book has helped recalibrate your mindset toward harmonious reintegration with nature's regenerative cycles.

While the surface-level goal was equipping you with the ability to achieve utility independence from extractive industrial models, the deeper purpose was designed to introduce you to something more than just residential self-sufficiency alone.

The true value embedded within each chapter was our shared commitment to wholeheartedly reconnect human existence back to nature's perpetual abundance without abusing it. It was about rekindling the ancient respect for our role as caretakers of the environment, not just as selfish exploiters of resources.

Whether deconstructing the linear "take-make-waste" consumption mentality insulating us from nature's closed-loop wisdom, or reawakening innate talents for humbly shaping elemental materials through skilled handicrafts, each section aims to dismantle alienating industrial barriers severing our sense of ecological belonging.

Despite absorbing a pretty comprehensive amount of knowledge that spans everything from renewable construction, off-grid food production, community networking, and survival preparedness, perhaps this book's greatest takeaway involves demystifying one simple truth—True self-sufficiency is an infinite journey of personal growth and transcendence, not merely a final destination.

Much like organic cultivation itself, realizing holistic regenerative autonomy demands constant tending, evolving understandings, and infinite patience to nurture resilience over time. So while we conclude this book's pages together, your self-reliant awakening has really just begun. The real "coursework" now involves internalizing those sustainable mindsets, then continuously applying them across every unfolding life choice and unmapped opportunity from here on out. Striving for congruence between your deepest values and each mundane action.

Which self-reliant pathways will you boldly step into first? Will you finally birth that long-envisioned tiny home construction project featuring innovative plant-based biocomposites and recycled material foundations? Or does urban homesteading localized self-provisioning awaken entrepreneurial dreams of decentralized closed-loop permaculture operations meeting your community's needs? Perhaps resilience to you is the ability to create a community that embraces the same philosophies as you—to create, share, and give back.

No matter what drives your independence, the key is to stay on that self-reliant path for the long haul. While unmapped detours will undoubtedly test your patience periodically, those minor setbacks are paler than a frozen Christmas turkey compared to generational fulfillment. Embrace the challenges that come with personal growth and self-realization. Revel in seemingly messy "failures" too; they merely temporarily obscure deeper wisdom. Embracing your connection to nature and living in harmony with its regenerative cycles allows you to find true independence and purpose beyond personal victories and challenges.

Always remain humble, open students no matter what credentials or plateaus you achieve.

So keep bravely questioning dogmas, stretching comfort zones, and courageously experimenting at the risk of unlearning previous "mastery." No trail remains open without continual intentional upkeep, redefining those footpaths. Ultimately, achieving genuine self-reliance means skillfully reconnecting with our environment through nurturing regenerative awareness. It's about staying devoted to honoring the sacred harmony present in all life, guiding each decision and action thereafter. So, stay genuinely committed to that sustainable path. Only those who uphold ethical responsibilities as masters can fuel humanity's ecological empowerment revolution, reigniting our potential for brighter futures.

If any seedlings of self-reliant wisdom took root here, amplifying your life's journey even fractionally, I have one humble request, pay these lessons forward by nurturing others' callings. I'd be truly honored if you could leave an honest review or feedback as well. By spreading empowered resilience, humanity can beautifully reconnect with our beloved planet. Step out ahead with an abundance mentality. Let these pages dissolve any lingering, limiting industrial mind viruses constraining your dreams, and embrace well-earned self-reliant thriving!

At this juncture, the choice becomes yours alone—Will you boldly embody the role of a self-reliant master? Courageously nurturing harmonious homeostasis within Mother Nature's regenerative patterns despite any lingering fears or doubts?

Or will you default back into disconnected dogmas severing you from the primordial balances whispering ancient wisdom non-stop? Remember, this Earth is merely borrowed to us by our future generations.

The decision rests squarely within your heart's authenticity alone. I have faith that you'll make choices that amplify the greatest harmonies within yourself. We all desire the deep fulfillment that comes from actively caring for the environment in everything we do. Those who have experienced even a glimpse of this regenerative bliss understand that there's no better path to nourishing the soul.

It's by honoring ethical ecological duties that humanity rediscovers its true essence as cherished custodians, tasked with guiding our planet's magnificent story to its flourishing finale. You were born for nothing less than that starring role anyway.

REFERENCES

Access, I. P. (2022, October 18). *How to set up off-grid communications for first responders.* IP Access International. https://www.ipinternational.net/how-to-set-up-reliable-off-grid-communication-at-a-crisis-center/

Adamant, A. (2018, February 28). *Financing off grid property - Things to know.* Practical Self Reliance. https://practicalselfreliance.com/off-grid-financing/

Admin. (2024, March 22). *Sustainable off-grid living: A comprehensive guide to house plans and design.* Houseplanstory. https://houseplanstory.com/off-grid-housing-plans/

admin. (n.d.). *Ultimate guide to DIY off grid projects: Master sustainable living one project at a time.* Eco Living Vibes. Retrieved March 30, 2024, from https://www.ecolivingvibes.com/diy-off-grid-projects-ultimate-guide-130-ideas-to-master-eco-life/#google_vignette

Administrator. (n.d.). *Tips for making a home defense plan.* The Hub AZ. Retrieved April 2, 2024, from https://thehubaz.com/blog/tips-for-making-a-home-defense-plan/

Bartering 101: Understanding the trade. (n.d.). Mailchimp. https://mailchimp.com/resources/bartering/

Beginners guide to starting a self-sustaining farm. (2024, February 4). Worldpackers. https://www.worldpackers.com/articles/self-sustaining-farm

Bonsall, W. (n.d.). *Crop rotation in the garden.* Maine Organic. https://www.mofga.org/resources/gardening/crop-rotation-in-the-garden/

Bowen, P. (n.d.). *Living off the grid.* Bleecker Street. Retrieved April 2, 2024, from https://bleeckerstreetmedia.com/editorial/living-off-the-grid

Burton-Hughes, L. (2018, December 19). *Food preservation methods and guidance*. High Speed Training. https://araven.com/en/actualidad/blog/guide-to-food-preservation-methods/

Buying land for off-grid living: Tips and considerations. (n.d.). Communitylands. Retrieved March 30, 2024, from https://communitylands.com/communitychronicle/buying-land-for-off-grid-living:-tips-and-considerations

Canada. (n.d.). *Health and safety programs*. CCOHS. https://www.ccohs.ca/oshanswers/hsprograms/planning.html

Cao, L. (2023, December 5). *How does a trombe wall work?* ArchDaily. https://www.archdaily.com/946732/how-does-a-trombe-wall-work

Carter, C. (2021, July 2). *How did people tell time before clocks?* Jack Mason. https://jackmasonbrand.com/blogs/news/how-did-people-tell-time-before-clocks

CCTV installation guide: All you need to know. (n.d.). Safe and Sound Security. https://getsafeandsound.com/blog/security-camera-installation/

Chaple, G. (2008, March 10). *Learn the constellations*. Astronomy Magazine. https://www.astronomy.com/observing/learn-the-constellations/

Cherlinka, V. (2023, February 13). *Crop rotation: A way to boost yields*. Eos Data Analytics. https://eos.com/blog/crop-rotation/

Coffin, R. (2024). *Off grid solar for dummies: Updated for 2024*. Power Station NZ - off Grid Solar. https://powerstation.nz/off-grid-solar-for-dummies-2021/

COGAN, D., COONY, J., & RESANKOVA, D. (2020, May 4). *Financial resilience for off-grid solar is more important now than ever before*. World Bank Blogs. https://blogs.worldbank.org/en/energy/financial-resilience-grid-solar-more-important-now-ever

Conover, T. (2022, December 1). *The politics of independence: Living off-grid in the Colorado foothills*. Literary Hub. https://lithub.com/the-politics-of-independence-living-off-grid-in-the-colorado-foothills/

Considerations on how to live off the grid. (2023, April 7). Anker. https://www.

anker.com/blogs/outdoor/considerations-on-how-to-live-off-the-grid

Coutard, O., Bothereau, B., & Tarr, J. (2023). History (and stories) of off-grid technologies: a reappraisal. *Flux, 131*(1), 1–14. https://doi.org/10.3917/flux1.131.0001

Create your emergency plan in just 3 steps. (n.d.). American Red Cross. https://www.redcross.org/get-help/how-to-prepare-for-emergencies/make-a-plan.html

Creating & storing an emergency water supply. (2021, January 26). Centers for Disease Control and Prevention. https://www.cdc.gov/healthywater/emergency/creating-storing-emergency-water-supply.html

Creating defensible space for your home and property. (n.d.). Arcata Fire District. Retrieved April 2, 2024, from https://www.arcatafire.org/creating-defensible-space-for-your-home-and-property

Crop rotation for seed growers. (2020, December 14). Dan Brisebois. https://danbrisebois.com/2020/12/14/crop-rotation-for-seed-growers/

Cyrus. (2020, March 1). *The importance of the off-grid/homestead community.* Offgridmaker. https://offgridmaker.com/off-grid-mindset/the-importance-of-the-off-grid-homestead-community/

D, W. (2023, April 26). *Energy storage system for off-grid systems.* LinkedIn. https://www.linkedin.com/pulse/nergy-storage-system-off-grid-systems-winnie-deng/

Dean. (2024, February 7). *Growing your own food for off grid living.* BeamBound. https://beambound.com/growing-your-own-food-for-off-grid-living/

Defensive landscaping: Using plants for protection. (2024, January 25). Security.org. https://www.security.org/blog/defensive-landscaping-using-plants-for-protection/

Developing a disaster communication plan. (2023, January 18). Tulane University. https://publichealth.tulane.edu/blog/developing-disaster-communication-plan/

Deziel, C. (2024, February 29). *Here's how to store water long-term for emergencies*. Family Handyman. https://www.familyhandyman.com/article/how-to-store-water-long-term/

Discovering off the grid communities: What you need to know and how to join them. (2023, October 11). Worldpackers. https://www.worldpackers.com/articles/off-the-grid-communities

Ellis, J. (2024, February 16). *The history of the power grid in the United States*. Landgate. https://www.landgate.com/news/the-history-of-the-power-grid-in-the-united-states

Emergency preparedness tips. (n.d.). Nationwide. https://www.nationwide.com/lc/resources/emergency-preparedness/articles/catastrophe-preparation

Emergency response plan. (2023, December 22). Ready.gov. https://www.ready.gov/business/emergency-plans/emergency-response-plan

Environmental issues news. (2024, April 1). ScienceDaily. https://www.sciencedaily.com/news/earth_climate/environmental_issues/

Evans, S. (2018, September 27). *Offgrid communities: using renewable energy to live independently*. Power Technology. https://www.power-technology.com/features/offgrid-communities-renewable-energy/

Everett, W. (2023, September 5). *Answering some questions about off-grid living*. Insteading. https://insteading.com/blog/questions-about-off-grid-living/

Everything you need to know about going off-grid. (n.d.). Plug In: Retrieved April 2, 2024, from https://www.energytechguide.com.au/everything-you-need-to-know-about-going-off-grid

Everything you need to know about going off-grid with solar. (n.d.). AltE. https://www.altestore.com/diy-solar-resources/everything-you-need-to-know-about-going-off-grid-with-solar/

Friedell, D. (2022, June 19). *"Off grid" living no longer means far away from others*. VOA. https://learningenglish.voanews.com/a/

off-grid-living-no-longer-means-far-away-from-others/6615990.html

Glines, A. (2024, February 15). *Cultural and historical perspectives on off-grid living*. Medium. https://imallenglines.medium.com/cultural-and-historical-perspectives-on-off-grid-living-afd6ef0e2182

Gordon, T. (n.d.). *Aquaponics gardening: A beginner's guide to building your own aquaponic garden*. Everand. Retrieved March 30, 2024, from https://www.everand.com/book/497586334/Aquaponics-Gardening-A-Beginner-s-Guide-to-Building-Your-Own-Aquaponic-Garden

Grazulis, D. (2021, August 5). *Financing an off-grid property with a land loan*. Life Lived Curiously. https://lifelivedcuriously.com/land-loan-homestead/

Hailey, L. (n.d.). *How to live off the grid: the ultimate guide for beginners (2023)*. EcoShack. https://ecoshack.com/how-to-live-off-the-grid/

Hall, B. (2023, March 3). *The benefits of off grid living for mental health & wellbeing*. Live off Grid. https://liveoffgrid.co.uk/the-benefits-of-off-grid-living-for-mental-health-wellbeing/

Harbour, S. (2021, January 23). *Off grid living mistakes I made & how to avoid them*. An off Grid Life. https://www.anoffgridlife.com/off-grid-living-mistakes/

Hayes, B. (n.d.). *15 DIY projects for preppers*. Urban Survival Site. https://urbansurvivalsite.com/diy-projects-preppers/

Hollaar, K. (2024, January 30). *Back to the origins: How off-grid living reawakens appreciation for life's essentials*. Medium. https://medium.com/@afkatja/back-to-the-origins-how-off-grid-living-reawakens-appreciation-for-lifes-essentials-35e76952b47a

Homemade water purifier. (n.d.). Extension.usu.edu

How to layout your self-sustaining permaculture farm. (2023, January 3). Our Wild Garden. https://ourwildgarden.com/how-to-layout-your-self-sustaining-permaculture-farm/

How to live off the grid with no money: A practical guide. (2023, September

1). Worldpackers. https://www.worldpackers.com/articles/how-to-live-off-the-grid-with-no-money

How to make compost at home. (2024, February 13). University of Maryland Extension. https://extension.umd.edu/resource/how-make-compost-home/

How to save water? (n.d.). KAS. Retrieved April 2, 2024, from https://kas.com.tr/en/blog/how-to-save-water/

Hutchinson, E. (n.d.). *How much would it cost to go off-grid with solar and wind power, including buying all equipment and paying for installation costs of both.* Quora. Retrieved April 2, 2024, from https://www.quora.com/How-much-would-it-cost-to-go-off-grid-with-solar-and-wind-power-including-buying-all-equipment-and-paying-for-installation-costs-of-both-installations

HydroponicTrash. (2022, July 15). *Recipes for an off-grid "I\internet."* Sunshine and Seedlings: A Newsletter by HydroponicTrash. https://anarchosolarpunk.substack.com/p/offgridinternet

Ink, O. (2016, March 27). *My off takes from off-grid philosophy.* Medium. https://medium.com/@OctopusInk/my-off-takes-from-off-grid-philosophy-a6c498f99d8f

Installing and maintaining a small wind electric system. (n.d.). Energy Saver. https://www.energy.gov/energysaver/installing-and-maintaining-small-wind-electric-system

Isolation vs community the pros & cons when living off the grid. (2023, June 1). Zero & Zen. https://www.zerozen.co.uk/isolation-vs-community-the-pros-cons-when-living-off-the-grid/

Jack, J. I. C. (2024, March 26). *Unique home fortification tactics to punish an invader.* Skilled Survival. https://www.skilledsurvival.com/home-fortification-tips/

julinasmall. (2022, May 11). *Financing an off-grid or earth sheltered home.* Little Dog Ranch. https://www.littledogranch.com/post/financing-an-off-grid-or-earth-sheltered-home

Kammen, & Daniel. (2015, July 24). *SMART vILLAGES: New thinking for off-grid communities worldwide*. Rael. https://rael.berkeley.edu/publication/smart-villages-new-thinking-for-off-grid-communities-worldwide/

Kashima, Y. (2020). Cultural dynamics for sustainability: How can humanity craft cultures of sustainability? *Current Directions in Psychological Science, 29*(6), 538–544. https://doi.org/10.1177/0963721420949516

Kelly, H. (2023, February 25). *Would you swap suburbia for self-sufficiency? As families face sky-high mortgage rates and inflation, these couples reveal how they chose to live off the grid.* Mail Online. https://www.dailymail.co.uk/news/article-11780923/Self-sufficiency-movement-sees-rise-people-living-grid.html

Kenton, W. (n.d.). *Barter (or bartering) definition, uses, and example*. Investopedia. https://www.investopedia.com/terms/b/barter.asp

Krosofsky, A. (2020, December 23). *What is permaculture gardening? You can create a self-sufficient garden*. Green Matters. https://www.greenmatters.com/p/what-is-permaculture-gardening

Lance , J. (2020, October 1). *Living off-grid: Our micro hydro alternative energy system*. Insteading. https://insteading.com/blog/living-off-grid-micro-hydro-alternative-energy-system/

Learn Hub: Do it yourself solar. (n.d.). Unbound Solar. https://unboundsolar.com/solar-information/diy-solar

Lillie, P. (2023). Off-grid living for the normative society: Shifting perception and perspectives by design. *University of Massachusetts Amherst*. https://doi.org/10.7275/35386782

Mack, E. (2023a, September 7). *Living off-grid comes with both savings and hidden expenses*. CNET. https://www.cnet.com/home/energy-and-utilities/living-off-grid-comes-with-both-savings-and-hidden-expenses/

Mack, E. (2023b, September 7). *Three years in, the biggest benefits and struggles of life off-grid surprise me*. CNET. https://www.cnet.com/home/energy-and-utilities/three-years-in-the-biggest-benefits-and-struggles-of-life-off-grid-surprise-me/

Mack-Heller, C. (2021, April 22). *5 Cross-sector collaboration examples for conservation and climate change impact*. Resonance Global. https://www.resonanceglobal.com/blog/5-cross-sector-collaboration-examples-for-conservation-and-climate-change-impact

Make a first aid kit. (n.d.). American Red Cross. https://www.redcross.org/get-help/how-to-prepare-for-emergencies/anatomy-of-a-first-aid-kit.html

Masonry heater. (n.d.). Wikipedia. Retrieved March 29, 2024, from https://en.wikipedia.org/wiki/Masonry_heater

Matthews, K. (2022, January 7). *How to make a DIY water filtration system using sand or gravel*. Mother Earth News. https://www.motherearthnews.com/diy/how-to-make-a-diy-water-filtration-system/

McGrath, M. K. (n.d.). *Legal hurdles in real estate: Navigating zoning and building codes*. Northspyre. https://www.northspyre.com/blog/legal-hurdles-in-real-estate-navigating-zoning-and-building-codes

Melchiore, R., & Melchiore, R. & J. (2023, October 23). *Self-sufficiency versus off-grid, is there a difference?* Off Grid and Free: My Path to the Wilderness. http://inthewilderness.net/2023/10/23/self-sufficiency-versus-off-grid-is-there-a-difference/#:~:text=People%20off%2Dgrid%20make%20their

Member, B. F. T. (2020, April 24). *The legal constraints of off-grid building*. Builderfinance. https://www.builderfinance.com/blog/legal-constraints-off-grid-building

NHS Choices. (n.d.). *What should I keep in my first aid kit?* NHS. https://www.nhs.uk/common-health-questions/accidents-first-aid-and-treatments/what-should-i-keep-in-my-first-aid-kit/

Norman Wilfred Desrosier, & R. Paul Singh. (2024). Food preservation. In *Encyclopædia Britannica*. https://www.britannica.com/topic/food-preservation

Off-grid energy companies and financial institutions - Need and options for collaboration. (n.d.). Energypedia. Retrieved April 2, 2024, from https://energypedia.info/wiki/Off-grid_Energy_Companies_and_Financial_Institutions_-_Need_

and_Options_for_Collaboration

Off-grid energy storage systems. (n.d.). Energian. https://www.energian.co.uk/collections/off-grid-all-in-one-solar-panel-kits

Off-grid living 101: Everything you need to know | Ultimate guide 2023. (2023, August 3). UGREEN. https://www.ugreen.com/blogs/off-grid-living/off-grid-living-101-ultimate-guide

Panish, D. (2020, December 21). *Automation goes off-grid.* Processing Magazine. https://www.processingmagazine.com/process-control-automation/article/21164839/automation-goes-off-grid

Ponce, L. (2021, July 19). *Things to consider before living off-the-grid.* THRIVE. https://blog.strive2thrive.earth/off-the-grid-living-considerations/

Preparedness tips. (n.d.). DEM State of Nevada. https://dem.nv.gov/preparedness/Preparedness_Tips/

Protect your property from wildfires with defensible spaces. (2023, July 10). Tidal Basin. https://www.tidalbasingroup.com/protect-your-property-from-wildfires-with-defensible-spaces/

published, R. C. (2022, July 15). *Sustainable garden ideas – 28 ways to create an eco-friendly garden.* Homes and Gardens. https://www.homesandgardens.com/gardens/create-an-eco-friendly-garden-220348

Raider, H. (2020, November 13). *How to strengthen your network when you're just starting out.* Harvard Business Review. https://hbr.org/2020/11/how-to-strengthen-your-network-when-youre-just-starting-out

Rainwater harvesting 101. (n.d.). Innovative Water Solutions LLC. https://www.watercache.com/education/rainwater-harvesting-101

Reardon, C. (n.d.). *Thermal mass.* Your Home. https://www.yourhome.gov.au/passive-design/thermal-mass

Robinson, D. (2024, January 3). *15 Biggest environmental problems of 2024.* Earth.org ; EARTH.ORG. https://earth.org/

the-biggest-environmental-problems-of-our-lifetime/

Russian stove. (n.d.). Wikipedia. Retrieved March 29, 2024, from https://en.wikipedia.org/wiki/Russian_stove#:~:text=A%20Russian%20stove%20is%20designed

Savitha, G., Aashish Ramana, S., & Jain, K. (2023, March 31). *Advanced security systems for home surveillance.* IEEE Xplore. https://ieeexplore.ieee.org/document/10058683

Self-sustaining garden - A garden that cares for itself. (n.d.). Smiling Gardener. Retrieved April 2, 2024, from https://www.smilinggardener.com/organic-gardening-tips/self-sustaining-garden/

Should I go off-grid or get a grid connection? (n.d.). OffGrid Australia. Retrieved March 30, 2024, from https://offgridaustralia.com.au/your-guide-to-living-with-off-grid-power/

Simoneau, T. (n.d.). *Life and design off grid.* Habitus. Retrieved April 2, 2024, from https://www.habitusliving.com/projects/architecture-off-grid-homes

Sipe, C. (2023, February 9). *Four strategies to prepare for an emergency.* EHS Daily Advisor. https://ehsdailyadvisor.com/2023/02/four-strategies-to-prepare-for-an-emergency/

So, you're thinking about going off the grid. (2021, April). Ecoliv. https://ecoliv.com.au/blog/going-off-the-grid

Solano, E. (n.d.). *How to build a DIY solar panel ground mount.* Tinktube. Retrieved April 8, 2024, from https://tinktube.com/free-plans/diy-solar-panel-ground-mount/

Solar energy & smart home automation: Sustainable living. (2024, February 15). Solar Emporium. https://solaremporium.com.au/solar-energy-smart-home-automation-sustainable-living/

Solar tips & tutorials. (n.d.). Today's Homeowner. Retrieved April 2, 2024, from https://todayshomeowner.com/solar/tips-and-tutorials/

Staff. (2023, February 6). *Off-grid living is expected to put a strain on utilities.* Vision Monday. https://www.visionmonday.com/business/research-and-stats/article/offgrid-living-is-expected-to-put-a-strain-on-utilities/#:~:text=The%20study%20found%20that%2012

Tackling threats that impact the earth. (n.d.). World Wildlife Fund. https://www.worldwildlife.org/threats

Team, O. G. W. (2023, March 19). *Off grid septic system and wastewater treatment.* Off Grid World. https://offgridworld.com/off-grid-septic-system/

Telling the time with the Sun. (2018, September 26). Science Museum. https://www.sciencemuseum.org.uk/objects-and-stories/telling-time-sun

The Bulworthy Project. (2015, August 3). *How to get planning permission for an off-grid self-build home.* Lowimpact.org. https://www.lowimpact.org/posts/how-to-get-planning-permission-for-a-rural-off-grid-self-build-home

The legality of off-grid living: A comprehensive guide. (2023, June 28). Zero & Zen. https://www.zerozen.co.uk/the-legality-of-off-grid-living-a-comprehensive-guide/

The role of advanced surveillance in modern security strategies. (n.d.). Pavion. Retrieved March 30, 2024, from https://pavion.com/resource/the-role-of-advanced-surveillance-in-modern-security-strategies/?subpage=articles

The steps to an efficient rainwater harvesting system. (n.d.). UltraTech Cement. Retrieved April 2, 2024, from https://www.ultratechcement.com/for-homebuilders/home-building-explained-single/descriptive-articles/the-steps-to-an-efficient-rainwater-harvesting-system

Thomas, J. (2021, June 19). *How to make compost the easy way.* Homesteading Family. https://homesteadingfamily.com/how-to-make-compost-the-easy-way/

Thomas, S. (n.d.). *54 DIY projects for preppers.* The Survival Journal. Retrieved March 30, 2024, from https://thesurvivaljournal.com/prepper-projects/

tlankford01. (n.d.). *Off-Grid open community mesh network.* Hackaday.Io.

Retrieved April 2, 2024, from https://hackaday.io/project/1482-off-grid-o-pen-community-mesh-network

Top 10 advantages and drawbacks of off-grid living. (2022, October 14). Discount Lots. https://discountlots.com/off-grid-living-advantages-and-drawbacks/

Tyrell, F. (2020, August 19). *10 Low-cost home defense tactics you can implement today.* Secrets of Survival. https://secretsofsurvival.com/10-low-cost-home-de-fense-tactics/

Understanding the different types of security systems & their uses. (n.d.). Lodge Service. https://lodgeservice.com/understanding-the-different-types-of-securi-ty-systems-their-uses/

Up North and Off Grid. (2022, January 8). *#0 The Goals.* Medium. https://medium.com/@upnorthandoffgrid/0-the-goals-e7233a2eb57

Vernazza , L. (2023, April 13). *Building an off-grid house: What you need to know.* The Plan Collection. https://www.theplancollection.com/blog/building-an-off-grid-house

Water quality & distribution activity for kids. (n.d.). Generation Genius. https://www.generationgenius.com/activities/water-quality-and-distribution-activi-ty-for-kids/

Weber, M. (n.d.). *Living off the grid: The complete guide for beginners.* Next-StepLiving. Retrieved March 30, 2024, from https://www.nextstepliving.com/living-off-grid

What are video surveillance systems and how do they work? (2024, March 3). Mammoth Security Inc. https://mammothsecurity.com/blog/how-video-sur-veillance-systems-work

What to put in a first aid kit. (n.d.). St John Ambulance. https://www.sja.org.uk/get-advice/i-need-to-know/what-to-put-in-a-first-aid-kit/

What's an "interactive energy grid"? (n.d.). Automation & Entertainment Inc. Retrieved March 30, 2024, from https://www.automationandentertainment.

com/residential-efficient-power-systems

Why take a first aid class? Save time and money. Personalized, flexible learning. (n.d.). Red Cross. https://www.redcross.org/take-a-class/first-aid

Wikipedia Contributors. (2018, November 29). *Constellation.* Wikipedia; Wikimedia Foundation. https://en.wikipedia.org/wiki/Constellation

Willsher, I. (2022, December 12). *The pros and cons of off-grid living.* Utopia. https://utopia.org/guide/the-pros-and-cons-of-off-grid-living/

Wu, B. (2023, November 14). *What is energy storage: A comprehensive guide.* LinkedIn. https://www.linkedin.com/pulse/what-energy-storage-comprehensive-guide-ben-wu-27qkc/

Zafar |, S. (2023, November 2). *Eco-friendly and sustainable security options for your entire home.* EcoMENA. https://www.ecomena.org/eco-friendly-sustainable-security-options/

Zenghelis, D. (2021, October 11). *What are the likely costs of the transition to a sustainable economy?* Economics Observatory. https://www.economicsobservatory.com/what-are-the-likely-costs-of-the-transition-to-a-sustainable-economy

Image References

Casap, E. (2016). Metal Backgrounds Tomato Garden [Image]. In *Unsplash.* https://unsplash.com/photos/bowl-of-tomatoes-served-on-person-hand-qgH-GDbbSNm8

Chalcraft, C. (2019). [Image]. In *Unsplash.* https://unsplash.com/photos/brown-handled-ax-on-tree-trunk-LVDi4ldYCVU

Chuangchoem, A. (2017). [Image]. In *Pexels.* https://www.pexels.com/photo/water-droplets-on-green-leaf-347925/

Follow, P. (2016a). [Image]. In *Pexels.* https://www.pexels.com/photo/house-floor-plan-271667/

Follow, P. (2016b). [Image]. In *Pexels.* https://www.pexels.com/photo/

silver-security-camera-207574/

Follow, P. (2017a). [Image]. In *Pexels*. https://www.pexels.com/photo/blue-so-lar-panel-board-356036/

Follow, P. (2017b). [Image]. In *Pexels*. https://www.pexels.com/photo/group-of-people-holding-arms-461049/

Henderson, M. (2020). Money backgrounds [Image]. In *Unsplash*. https://un-splash.com/photos/green-plant-in-clear-glass-cup-SoT4-mZhyhE

Lastovich, T. (n.d.). [Image]. In *Pexels*. Retrieved January 18, 2018, from ht-tps://www.pexels.com/photo/photo-of-a-man-walking-on-boardwalk-808466/

McClung, J. (2018). My oldest daughter holding a bundle of swiss chard. [Ima-ge]. In *Unsplash*. https://unsplash.com/photos/person-holding-bok-choy-0aC-3Gt7jF4

Nilov, M. (2021). [Image]. In *Pexels*. https://www.pexels.com/photo/close-up-of-a-bag-with-emergency-aid-kit-8943184/

Owen, J. (2021). [Image]. In *Unsplash*. https://unsplash.com/photos/brown-wo-oden-house-near-green-tree-during-daytime-ZdEQ_JKizsU

Rakozy, G. (2015). Beautiful Pictures & Images [Image]. In *Unsplash*. https://unsplash.com/photos/silhouette-photography-of-person-oMpAz-DN-9I

RDNE Stock project. (2021). [Image]. In *Pexels*. https://www.pexels.com/pho-to/overhead-shot-of-a-paper-with-graphs-and-charts-7947663/

Rozetsky, A. (2017). Earth Images & Pictures [Image]. In *Unsplash*. https://unsplash.com/photos/brown-grain-field--c0PJUAtpSo

Schulte, M. (2019). Lensball on a rock at the "Dreimühlen" - waterfall in Ger-many [Image]. In *Unsplash*. https://unsplash.com/photos/clear-glass-ball-on-brown-rock-n5qzgURyBEI

Spiske, M. (2016). Rearing tomatoes for self-support [Image]. In *Unsplash*. https://unsplash.com/photos/selective-focus-photo-of-plant-spouts-vrbZVyX2k4I

Tosta, L. (2017). [Image]. In *Unsplash*. https://unsplash.com/photos/tilt-shift-lens-photography-of-black-steel-faucet-SVeCm5KF_ho

Wilson, L. (2017). Home grown greenhouse tomatoes, Northumberland [Image]. In *Unsplash*. https://unsplash.com/photos/person-holding-green-and-red-tomatoes-hkN0yoPrpM4

Woodhead, J. (2017). This photo was taken looking up at a classic Nordic house in Reykjavik, contrasted against the winter sky [Image]. In *Unsplash*. https://unsplash.com/photos/closed-window-tfEoT92EYlo

Tosta, L. (2017). [Image]. In Unsplash. https://unsplash.com/photos/tilt-shift-lens-photography-of-black-steel-faucet-SVeCm5KF_ho

Wilson, L. (2017). Home grown greenhouse tomatoes, Northumberland [Image]. In Unsplash. https://unsplash.com/photos/person-holding-green-and-red-tomatoes-hkN0yoPrpM4

Woodhead, J. (2017). This photo was taken looking up at a classic Nordic house in Reykjavik, contrasted against the winter sky [Image]. In Unsplash. https://unsplash.com/photos/closed-window-tfEoT92EYlo

Made in United States
Troutdale, OR
08/28/2024

22401328R00095